Implementing CI/CD Using Azure Pipelines

Manage and automate the secure flexible deployment of applications using real-world use cases

Piti Champeethong

Roberto Mardeni

<packt>

BIRMINGHAM—MUMBAI

Implementing CI/CD Using Azure Pipelines

Group Product Manager: Preet Ahuja

Publishing Product Manager: Surbhi Suman

Book Project Manager: Deeksha Thakkar

Senior Editor: Shruti Menon

Technical Editor: Yash Bhanushali

Copy Editor: Safis Editing

Proofreader: Safis Editing

Indexer: Tejal Daruwale Soni

Production Designer: Prashant Ghare

DevRel Marketing Coordinator: Rohan Dobhal

Senior DevRel Marketing Coordinator: Linda Pearlson

First published: November 2023

Production reference: 1301123

Published by
Packt Publishing Ltd.
Grosvenor House
11 St Paul's Square
Birmingham
B3 1RB, UK

ISBN 978-1-80461-249-1

www.packtpub.com

To my beautiful wife, Promyok, for always supporting me and challenging me to do better.

– Piti Champeethong

To my beautiful and incredibly talented wife, Marirose, for supporting me, always challenging me, and encouraging me to improve myself. To my kids, Joseph and Lamysse, for being the inspiration to never stop learning and always wanting to provide more.

– Roberto Mardeni

Contributors

About the authors

Piti Champeethong is a senior consulting engineer at MongoDB, Singapore. He has been a part of the software development community for over 20 years and specializes in database application development and CI/CD implementation. He is a Microsoft Certified Trainer, lecturer, and community leader. He lives in Thailand and engages with Thai tech communities, such as the MongoDB Thailand User Group and the Thailand .NET community by speaking at various events. He has made significant contributions to public repositories on .NET and MongoDB technologies, supporting the growth of the Thai tech community.

I want to thank all the people who have been close to me and supported me, especially my wife and family. I would also like to thank the Packt team for their efforts and support.

Roberto Mardeni has been working in the IT industry for over two decades. He specializes in software development, architecture, and DevSecOps practices around many different application platforms and CI/CD tools, typically focusing on Microsoft .NET and other open source programming languages. He has been working since 2017 as an application innovation specialist on the enterprise sales side at Microsoft, supporting their cloud business. He lives in the United States and helps some of the largest independent software vendors to adopt the Azure platform. This is his first foray into technical writing, but he has contributed to the open source community in different GitHub public repositories of his own and contributed to others as well.

First and foremost, I would like to thank my loving and patient wife and family for their continued support, patience, and encouragement throughout the long process of writing this book. Thanks also to the Packt team for their invitation to collaborate on the writing of this book.

About the reviewers

James Wasson, is a technology leader at Heartland Payments Systems and a global ambassador for the DevOps Institute. He has over 15 years of expertise in technology and building large-scale organizations. Currently, he leads a global team of cloud engineers and site reliability engineers to support all Heartland SaaS products. He is dedicated to building beautiful people, character, and software, in that order.

Thank you to my wife, Deanna, for all her support in our lives, and for those I get the chance to serve.

Aditya Sharma, a 16-year veteran of Microsoft technologies and a skilled Azure cloud engineer, excels in global project success. His track record showcases cost-effective cloud solutions, enhancing application performance, and upholding security in cloud environments. Aditya boasts Microsoft certifications, including Azure Developer Associate and DevOps Engineer Expert, underscoring his tech commitment. Proficient in Azure services such as App Service and Logic Apps, Aditya optimizes performance and cost efficiency. His passion for innovation and problem-solving sets him apart as a valuable team asset.

I would like to extend my heartfelt gratitude to my family and friends for their unwavering understanding and support throughout my journey. As a highly skilled .NET technologies project manager, I have had the privilege of leading an exceptional team of professionals, and it is their unwavering dedication, expertise, and collaborative spirit that have driven our project to fruition.

Table of Contents

3

4

Part 2: Azure Pipelines in Action

5

6

7

Monitoring Azure Pipelines   145

8

Provisioning Infrastructure Using Infrastructure as Code   171

Part 3: CI/CD for Real-World Scenarios

9

Implementing CI/CD for Azure Services   203

10

Implementing CI/CD for AWS   237

11

Automating CI/CD for Cross-Mobile Applications by Using Flutter   261

12

Navigating Common Pitfalls and Future Trends in Azure Pipelines   285

Index   297

Other Books You May Enjoy   304

Preface

Continuous integration (**CI**) is the practice of using automated tools to compile automatically and continuously, based on changes made by developers to source code, and packaging and testing them to ensure they are stable and able to provide the expected functionality.

Continuous delivery (**CD**) uses the artifacts created via CI and deploys those applications without human intervention to end users, ensuring these are always updated with the latest versions and, in most cases, performing additional verifications in multiple environments before reaching the final end users.

All of this is possible by taking advantage of Azure Pipelines, one of the leading platforms to support all aspects of the **software development life cycle** (**SDLC**); however, in this book, we will focus on the CI/CD aspects and how to take advantage of the features available to effect the different task automation options under the DevOps umbrella, even touching on the DevSecOps aspects.

In this book, we provide you with the tools to get started, by learning the basic concepts and building from there to more complex scenarios. This will lead to end-to-end scenarios that enable your software development teams with the tools to automate, every step of the way, the delivery of applications, from source code to a running platform.

Who this book is for

From beginners to the most advanced users, anybody looking to better understand how to take advantage of Azure Pipelines can benefit from this book.

The three main personas who are the target audience of this content are as follows:

- **Software developers**: They will learn how they can automatically build and deploy their software products, regardless of the target platform, at very early stages to speed up the SDLC
- **DevOps engineers**: They will learn how Azure Pipelines can support any automation need, no matter the stage in the process, infusing quality and checks every step of the way
- **Security engineers**: They will learn how they can integrate their tools in the CI/CD process to enforce security and quality at the beginning of the build and deployment process

What this book covers

Chapter 1, Understanding Azure Pipelines, provides an introduction to CI/CD, Azure DevOps, Azure Pipelines, and its components. It explains why an Azure pipeline is the right choice for certain cases, introduces other services under Azure DevOps such as Azure Repos, and guides you through setting up a new project, setting up a self-hosted agent, preparing a pipeline environment, and configuring agent pools and deployment groups.

Chapter 2, Creating Build Pipelines, teaches you how to create and manage pipelines, stages, jobs, tasks, triggers, and artifacts in Azure DevOps, as well as about running pipelines after code pushes to Azure Repos.

Chapter 3, Setting Variables, Environments, Approvals, and Checks, covers the creation of service connections, variable groups, secret files, and release pipelines in Azure DevOps. It also explains setting up service accounts for Azure Repos and GitHub connections. Additionally, you will learn about securely storing secret keys and using environments, with approvals and checks for stage progression control.

Chapter 4, Extending Advanced Azure Pipelines Using YAML, helps you to understand how to use YAML to create a build and release pipeline. It discusses in detail what the YAML syntax is to create stages, jobs, and tasks for web application deployment.

Chapter 5, Implementing the Build Pipeline Using Deployment Tasks, explores how to create and reuse a build task for the building process. This chapter covers the popular Node.js, NPM, .NET, Docker, and SQL Server deployment tasks using the YAML syntax.

Chapter 6, Integrating Testing, Security Tasks, and Other Tools, helps you understand how the extensibility of Azure Pipelines with other tools works. This chapter covers the popular tools SonarQube for code analysis and Jenkins for artifacts.

Chapter 7, Monitoring Azure Pipelines, teaches you how to monitor Azure Pipelines and related tasks, such as build tasks, deployment tasks, and pipeline agents. You will also learn how to build monitoring into pipelines to determine whether deployments improve or degrade the quality of a system.

Chapter 8, Provisioning Infrastructure Using Infrastructure as Code, examines how to create and reuse deployment tasks for the **Infrastructure as Code** (**IaC**) process. This chapter covers the popular IaC tools Terraform, Azure Bicep, and an ARM template using YAML syntax.

Chapter 9, Implementing CI/CD for Azure Services, shows you how to create the YAML and pipelines for Azure service deployment. You will learn how to set up and deploy applications on **Azure App Service**, **Azure Kubernetes Service** (**AKS**), **Azure Container Apps**, and **Azure Container Instances** (**ACI**).

Chapter 10, Implementing CI/CD for AWS, explores how to create YAML and pipelines to deploy containerized applications on different services such as **AWS Lightsail**, **Elastic Kubernetes Service** (**EKS**), and **Elastic Container Service** (**ECS**).

Chapter 11, Automating CI/CD for Cross-Mobile Applications by Using Flutter, dives into how to create a pipeline using YAML to automate the CI/CD of a mobile application build and release process. You will also learn how to implement YAML pipelines to deploy Flutter on Apple TestFlight and the Google Play Console, staging the environment of an end-to-end process.

Chapter 12, Navigating Common Pitfalls and Future Trends in Azure Pipelines, teaches you about common mistakes and suggests how to avoid them. This chapter also looks at potential future trends in Azure Pipelines.

To get the most out of this book

You will need to have a basic understanding of building and deploying applications with automation; however, this book will walk you through how this is done in Azure Pipelines. Each chapter has specific technical requirements.

Software/hardware covered in the book	Operating system requirements
Docker	Windows, Linux, or macOS
Visual Studio Code	Windows, Linux, or macOS

If you are using the digital version of this book, we advise you to type the code yourself or access the code from the book's GitHub repository (a link is available in the next section). Doing so will help you avoid any potential errors related to the copying and pasting of code.

Download the example code files

You can download the example code files for this book from GitHub at `https://github.com/PacktPublishing/Implementing-CI-CD-Using-Azure-Pipelines`. If there's an update to the code, it will be updated in the GitHub repository.

We also have other code bundles from our rich catalog of books and videos available at `https://github.com/PacktPublishing/`. Check them out!

Conventions used

There are a number of text conventions used throughout this book.

`Code in text`: Indicates code words in text, database table names, folder names, filenames, file extensions, pathnames, dummy URLs, user input, and Twitter handles. Here is an example: Add the following basic script – `echo "Hello Second Task on Linux"`.

A block of code is set as follows:

```
{
  "$schema": "https://schema.management.azure.com/schemas/2019-04-01/
deploymentTemplate.json#",
  "contentVersion": "",
  "apiProfile": "",
  "parameters": {   },
  "variables": {   },
  "functions": [   ],
  "resources": [   ],
  "outputs": {   }
}
```

Any command-line input or output is written as follows:

```
$id=az ad sp list –display-name azure-pipelines –query "[].id" -o tsv
```

Bold: Indicates a new term, an important word, or words that you see on screen. For instance, words in menus or dialog boxes appear in **bold**. Here is an example: "Start by clicking the **Environments** option under **Pipelines** in the main menu."

> **Tips or important notes**
> Appear like this.

Get in touch

Feedback from our readers is always welcome.

General feedback: If you have questions about any aspect of this book, email us at customercare@ packtpub.com and mention the book title in the subject of your message.

Errata: Although we have taken every care to ensure the accuracy of our content, mistakes do happen. If you have found a mistake in this book, we would be grateful if you would report this to us. Please visit www.packtpub.com/support/errata and fill in the form.

Piracy: If you come across any illegal copies of our works in any form on the internet, we would be grateful if you would provide us with the location address or website name. Please contact us at copyright@packtpub.com with a link to the material.

If you are interested in becoming an author: If there is a topic that you have expertise in and you are interested in either writing or contributing to a book, please visit authors.packtpub.com.

Share Your Thoughts

Once you've read *Implementing CI/CD Using Azure Pipelines*, we'd love to hear your thoughts! Scan the QR code below to go straight to the Amazon review page for this book and share your feedback.

https://packt.link/r/1804612499

Your review is important to us and the tech community and will help us make sure we're delivering excellent quality content.

Download a free PDF copy of this book

Thanks for purchasing this book!

Do you like to read on the go but are unable to carry your print books everywhere?

Is your eBook purchase not compatible with the device of your choice?

Don't worry, now with every Packt book you get a DRM-free PDF version of that book at no cost.

Read anywhere, any place, on any device. Search, copy, and paste code from your favorite technical books directly into your application.

The perks don't stop there, you can get exclusive access to discounts, newsletters, and great free content in your inbox daily

Follow these simple steps to get the benefits:

1. Scan the QR code or visit the link below:

https://packt.link/free-ebook/978-1-80461-249-1

2. Submit your proof of purchase
3. That's it! We'll send your free PDF and other benefits to your email directly

Part 1: Getting Started with Azure Pipelines

This part will walk you through the basics of Azure Pipelines, help you understand its concepts, and show you how to get started quickly to implement automated build and deployment jobs.

This part has the following chapters:

1
Understanding Azure Pipelines

This book will be one of your favorite books in the Microsoft DevOps world as it provides a comprehensive guide to learning all about **Azure Pipelines** and will allow you to become an experienced Azure DevOps engineer. An Azure DevOps engineer is an individual who is responsible for designing and implementing **continuous integration and continuous deployment** (**CI/CD**) pipelines using the Azure Pipelines service, which is a component of **Azure DevOps**. Azure DevOps is a group of Microsoft services that help a project team achieve the project's goal.

In this chapter, you will be introduced to the CI/CD and Azure DevOps concepts in greater detail and will learn about the advantages of using Azure Pipelines to create CI/CD pipelines for the application deployment process. More specifically, in this chapter, you will learn about the following topics:

- What is CI/CD?
- Introducing Azure DevOps
- Introducing Azure Pipelines and its components
- Comparing Azure Pipelines with other CI/CD tools
- Setting up agent pools
- Creating a **personal access token** (**PAT**)
- Setting up and updating self-host agents
- Setting up deployment groups

Technical requirements

You can find the code for this chapter at `https://github.com/PacktPublishing/Implementing-CI-CD-Using-Azure-Pipelines/tree/main/ch01`.

What is CI/CD?

CI/CD is the workflow process for automation development and deployment that developers should know about to improve their skills.

CI is the workflow process for automating the process of building and testing code whenever a team member commits changes to Git, which is a form of version control that's run on a source control repository platform such as Azure Repos, GitHub, GitLab, and others. CI creates a modern culture for all developers to share their code, including unit tests, by merging all changes into a shared version control repository after finishing a small task. CI runs based on committing code triggers to grab the latest code from the shared version control repository to build, test, and validate any branch that they commit. Using CI allows you to rapidly discover error code issues and correct them to ensure all developer code is of good quality.

CD involves automating the process of building, testing, configuring, and deploying from the CI workflow process to specific environments, such as QA, staging, and production.

This workflow is illustrated in the following figure:

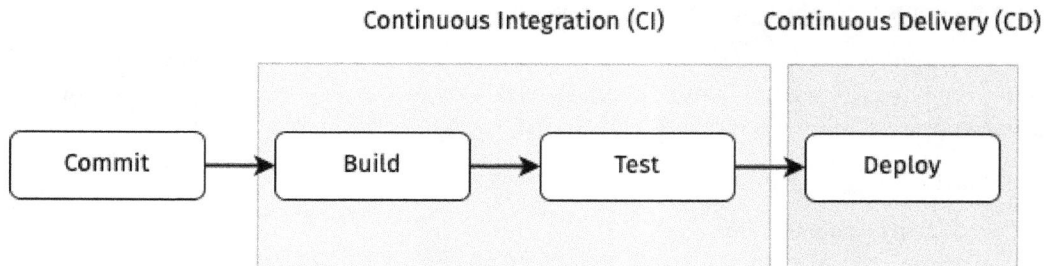

Continuous Integration (CI)   **Continuous Delivery (CD)**

Commit → Build → Test → Deploy

Figure 1.1 – CI/CD diagram

CI/CD reduces human error and the routine operation of the manual build, test, and deploy stages for any developer. It helps the developer focus only on application development.

This book will focus on the CI/CD tool **Azure Pipelines**, which is a comprehensive service for DevOps and a part of the ecosystem of services in **Azure DevOps**. Before we look at this further, let's introduce Azure DevOps.

Introducing Azure DevOps

Many CI/CD tools are used to support modern software development, such as Azure Pipelines, GitLab CI/CD, GitHub Actions, and Bitbucket Pipelines. One of the most widely used is Azure Pipelines; this is a part of **Azure DevOps**, which consists of the following five services:

- **Azure Boards** is an Azure DevOps sub-service that's used to track all tasks related to a project conveniently in one place. It is suitable for teamwork. It helps with collaboration because it supports Kanban boards, backlogs, team dashboards, and custom reporting, which can create a connection between the tasks and source version repositories such as GitHub or Azure Repos.

- **Azure Pipelines** is an Azure DevOps sub-service that's used to build, test, integrate, and deploy CI/CD processes. It helps reduce delivery errors and allows teams to focus solely on developing clean and readable code in software development; this service can be accessed through the Azure DevOps web portal (`https://dev.azure.com/{your-organization}`). This book will focus on this service due to this benefit.

- **Azure Repos** is an Azure DevOps sub-service for controlling the version of the source code. It is easy to manage code in one place. Easy maintenance can also help you define rules so that you can deploy code safely to desired environments, such as merge checks or static code analysis after the team creates the pull request. The examples in this book will use Azure Repos.

- **Azure Test Plans** is an Azure DevOps sub-service that helps test or quality assurance teams write use case scenarios to easily deliver the test results to the customer. The tester or quality assurance team creates **system integration testing** (**SIT**) and **user acceptance testing** (**UAT**) on Azure Test Plans. It can display test results as dashboard reports and include comments or feedback. Azure Test Plans also helps the team understand the test process of the project on the same page.

- **Azure Artifacts** is an Azure DevOps sub-service that enables developers to share and manage all their packages that result from building code in one place. Developers can publish packages to their feeds and share them within the same team, organization, and even publicly. Developers can also load the packages from different public repositories such as `https://www.nuget.org/` or `https://www.npmjs.com/`. Azure Artifacts also supports multiple package types, such as NuGet, npm, Python, Maven, and Universal Packages.

All these services fall under the umbrella of Azure DevOps, which covers the necessary development process for a project. You don't need to use additional services for development.

Introducing Azure Pipelines and its components

Azure Pipelines is a CI/CD platform for building, testing, and deploying your code to a live application. First, let's take a look at its key components.

Exploring the key components

There are some key concepts that you need to understand when creating an Azure pipeline:

- An **agent** is the software that runs a job within a server. It can be a Microsoft-hosted agent or a self-hosted agent.

- A **pipeline** is a workflow process for CI/CD for your application development. It can define your idea of how to build, test, integrate, and deploy your project.

- A **trigger** is an action that calls a pipeline to run.

- A **stage** is a flow of defined jobs in a pipeline, and each stage can have one or more jobs. The benefit of using a stage is that you can rerun job(s) under it. This means you do not need to rerun the whole pipeline. For example, let's say the developer creates a pipeline containing two stages: the **build stage** and the **deployment stage**. If the deployment stage fails, then they can only rerun the failed job under the deployment stage.

- A **job** is a group of one or more steps set in a stage. It is useful when you need to run a set of steps in a different operating system environment.

- A **step** can be a task or script and is the smallest piece of a pipeline:

  - A **task** is a pre-defined script that your idea can define.

  - A **script** is an action that uses the **command-line interface** (**CLI**), PowerShell, or Bash. It depends on the operating system agent that you choose for a job. For example, if you use a command line to run on a Linux agent, it will use a bash script. PowerShell runs on a macOS agent and will use a PowerShell core for cross-platform scripts.

- A **target** is a destination of the pipeline. It can be Azure Artifacts, an Azure resource service (such as Azure App Services, Azure Functions, Azure Container Apps, Azure Kubernetes Services, and so on), or invoke a REST API such as webhooks on Microsoft Teams.

Now, let's look at how these components interact with each other:

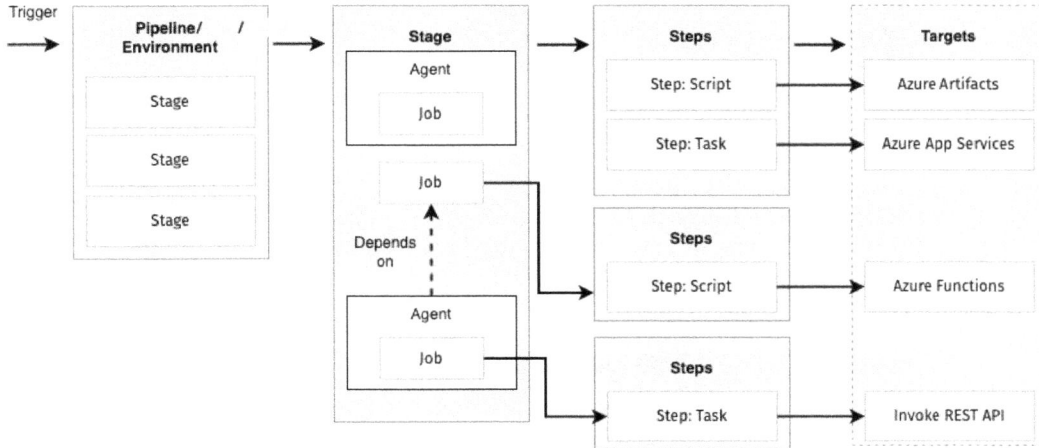

Figure 1.2 – Key components

This section described the meaning of and relationship between key objects. Before we take a more in-depth look at the different aspects of this platform, let's learn how we can start using it.

Signing up for Azure Pipelines

Two methods can be used for signing up:

- **Sign up with a Microsoft account**: To do this, complete the following steps:

  I. Go to https://azure.microsoft.com/en-us/services/devops/pipelines.

  II. Click on **Start free**.

  III. Log in with your Microsoft account.

  IV. Enter a name for your organization.

  V. You must always log in to your organization using https://dev.azure.com/{your-organization}.

- **Sign up with a GitHub account**: To do this, complete the following steps:

  I. Go to https://azure.microsoft.com/en-us/services/devops/pipelines.

  II. Click on **Start free with GitHub**.

  III. Log in with your GitHub account.

IV.   Enter a name for your organization.

V.   You must always log in to your organization using `https://dev.azure.com/{your-organization}`.

Once you've signed up for an Azure Pipelines account, you are ready to create a new project for building your code and release the built code to the live application.

Creating a new project

Creating a new project is the first step after you sign up, before creating any CI/CD pipeline. Upon creating a project, you can set project visibility:

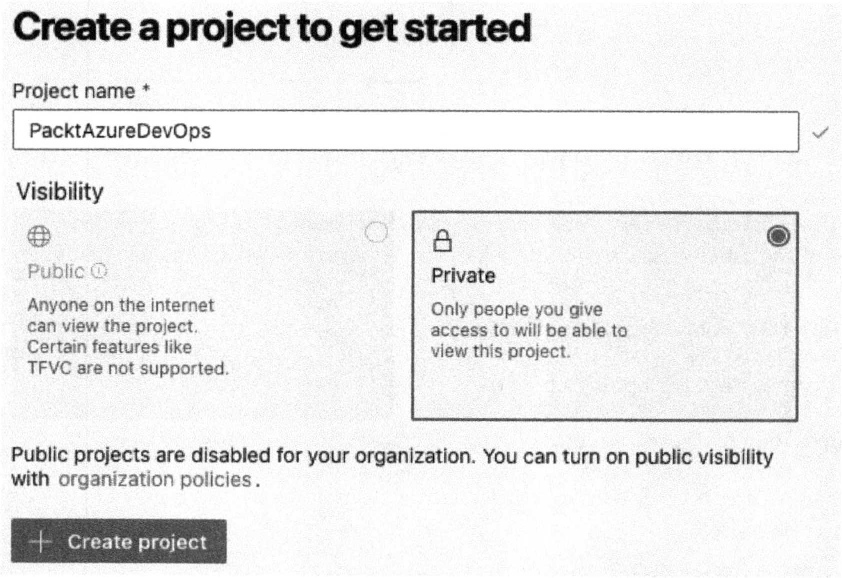

Figure 1.3 – Creating a new project

You can enter a project name and select **Visibility**, then click **Create project**.

Inviting team members

When you need to work with a team, you must add a new member by inviting one or more team members. Follow these steps to invite a team member:

1. Click on your project's name in the web portal and click **Project settings**:

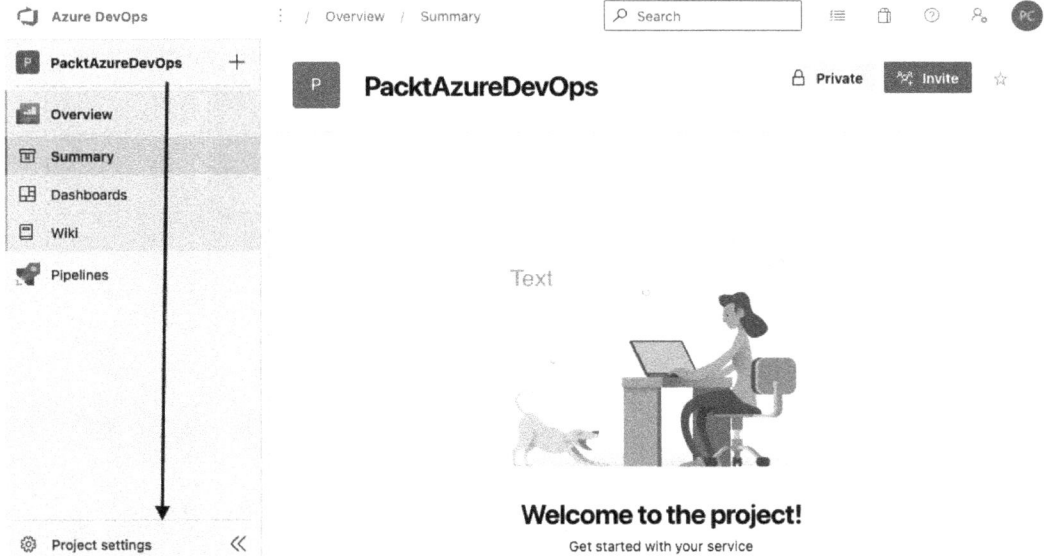

Figure 1.4 – Project settings

2.   Select **Teams | Add**:

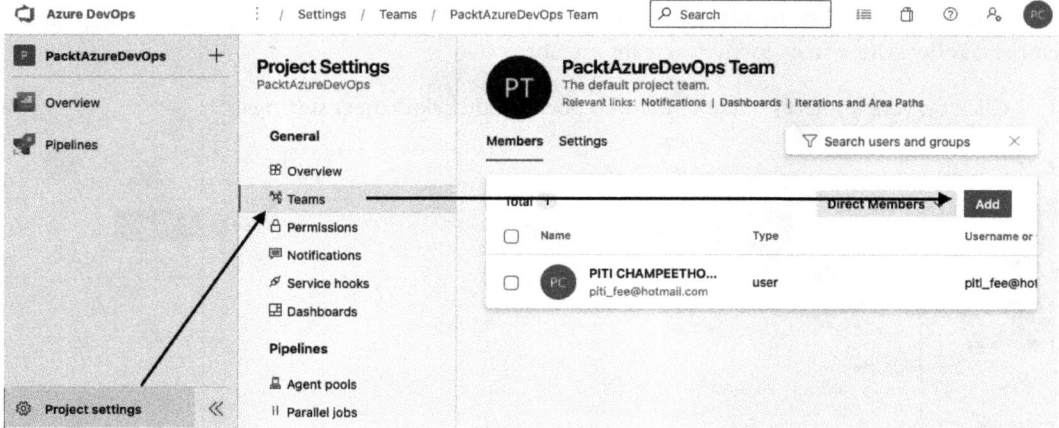

Figure 1.5 – Adding a new team member

3.   Enter the email addresses of your team members and click **Save**:

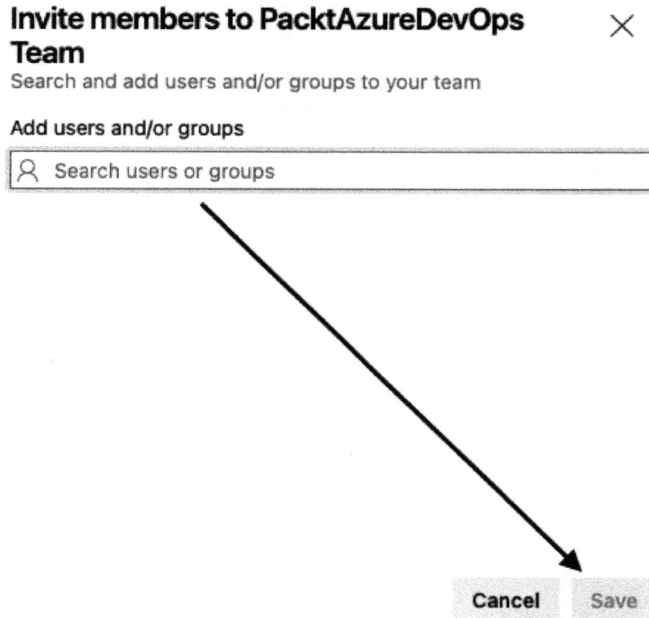

Figure 1.6 – Inviting team members

Now that you've invited your collaborators to join your project, let's delve into how we can start using this service.

Creating Azure pipelines

There are two ways to create Azure pipelines:

- Using the **Classic interface** (create Azure pipelines from the web Azure DevOps portal) while following these basic steps:

    I.   Configure Azure Pipelines to use your Azure Repos Git repository.

    II. Use Azure Pipelines to create and configure your build and release pipeline using drag and drop from the Azure DevOps portal.

    III. Push your code to your version control repository. The pipeline will be automatically initiated by the default trigger and the defined tasks will be executed.

- Using **YAML syntax** (create YAML files and commit them to the code repository) while following these basic steps:

    I.   Configure Azure Pipelines to use your Azure Repos Git repository.

    II. Edit your `azure-pipelines.yml` file by defining your custom build.

    III. Push your code to your version control repository. This action runs the default trigger.

Let's illustrate the Azure Pipelines YAML method for ease of understanding:

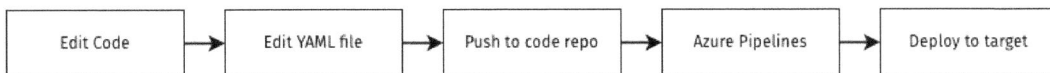

Figure 1.7 – Azure Pipelines YAML steps

There are different pipeline features available for both these methods, with some available for both and others only featuring in one. We'll look at these in detail in the next section.

Feature availability

Some pipeline features are only available when using the classic interface or YAML. The following table displays which features are available for which of these methods:

Feature	YAML	Classic	Description
Agents	Yes	Yes	To define the resource where the pipeline can run.
Approvals	Yes	Yes	To define the validation steps for additional checking before completing a deployment stage.
Artifacts	Yes	Yes	To define the library packages for publishing or consuming different package types.
Caching	Yes	Yes	To define an additional task to reduce the build time by allowing outputs or downloaded dependencies to store them on the agents and reuse them again.
Conditions	Yes	Yes	To define the specific conditions before running a job.
Container jobs	Yes	No	To define the specific jobs to run in a container.
Demands	Yes	Yes	To define the specific pipeline to ensure requirements are met before a pipeline stage is run.
Dependencies	Yes	Yes	To define specific requirements for validation before running the next job or stage.
Deployment groups	No	Yes	To define a logical group for the code that will be deployed to the target machines.
Deployment group jobs	Yes	Yes	To define a job to release to a deployment group.
Deployment jobs	Yes	No	To define the deployment steps.
Environment	Yes	No	To define a collection of resources targeted for deployment.
Gates	Yes	Yes	To support automatic collection and evaluation of external health signals before completing a release stage. Available with classic release only.
Jobs	Yes	Yes	To define the execution sequence of a set of steps.
Service connections	Yes	Yes	To define a connection to a remote service required to execute tasks in a job.
Service containers	Yes	No	To define a service that you can use to manage the life cycle of a containerized service.
Stages	Yes	Yes	To define flow jobs within a pipeline.
Task groups	No	Yes	To define a set of sequence tasks as a single reusable task.

Tasks	Yes	Yes	To define the building blocks that construct a pipeline.
Templates	Yes	No	To define reusable content, logic, and parameters.
Triggers	Yes	Yes	To define a specific event that causes a pipeline to run.
Variables	Yes	Yes	To define a value for data replacement and pass it to the pipeline.
Variable groups	Yes	Yes	To define the storage of values that you want to control and make available across multiple pipelines.

Table 1.1 – Pipeline features

Apart from these features, there are source version control repositories that Azure Pipelines can connect to. We'll look at these in detail in the next section.

Availability of source version control repositories

YAML pipelines only support some version control repositories. The following table displays which version control repositories can support which method:

Repository	YAML	Classic Interface
Azure Repos	Yes	Yes
GitHub	Yes	Yes
GitHub Enterprise Server	Yes	Yes
Bitbucket Cloud	Yes	Yes
Bitbucket Server	No	Yes
Subversion	No	Yes

Table 1.2 – Comparing repositories

In this section, we discussed all the available features of Azure Pipelines. In the next section, we will convert the key components of Azure Pipelines into a YAML structure to manage it better.

Understanding the YAML structure of Azure Pipelines

Usually, creating a file called `azure-pipelines.yml` will help you remember which YAML file is used for `azure-pipelines` in the source code repository. The basic Azure Pipelines YAML structure is as follows:

```
1    trigger:
2      - master
3
4    stages:
5      - stage: stage1
6        jobs:
7          - job: job1
8            pool:
9              vmImage: 'windows-latest'
10           steps:
11             - task: NuGetToolInstaller@1
12             - task: NuGetCommand@2
13               inputs:
14                 restoreSolution: 'mysolution.sln'
15             - script: echo Hello, world!
16               displayName: 'Run a one-line script'
17     - stage: stage2
18       dependsOn: stage1
19       jobs:
20         - job: importantJob
21           pool:
22             vmImage: 'windows-latest'
23           steps:
24             - pwsh: 'write-host "I do nothing"'
```

Figure 1.8 – azure-pipelines.yml file

The `azure-pipelines.yml` file in this example contains a typical structure:

- There are two stages, `stage1` and `stage2`, and each stage contains a `job` step.

- *Lines 1-2* show the pipeline runs when the developer pushes changes on the main (`master`) branch.

- *Lines 8-9* and *21-22* show the pipeline uses a Microsoft-hosted agent with the `windows-latest` operating system image.

- *Line 11* is a pre-created script for using the NuGet library. You can access this script in the `ch1` folder in this book's GitHub repository.

- *Line 12* is a pre-created script for using the NuGet command line.

- *Line 15* is a command line to run the echo command.

- *Line 24* is a PowerShell Core script that is cross-platform.

As you can see, the basic YAML structure is rather simple to understand. Once you've prepared the YAML file, you can see the status for running it. We will discuss this in the next section.

Viewing the Azure pipelines' status

The Azure pipelines' status is displayed on the Azure DevOps web portal under the running pipeline:

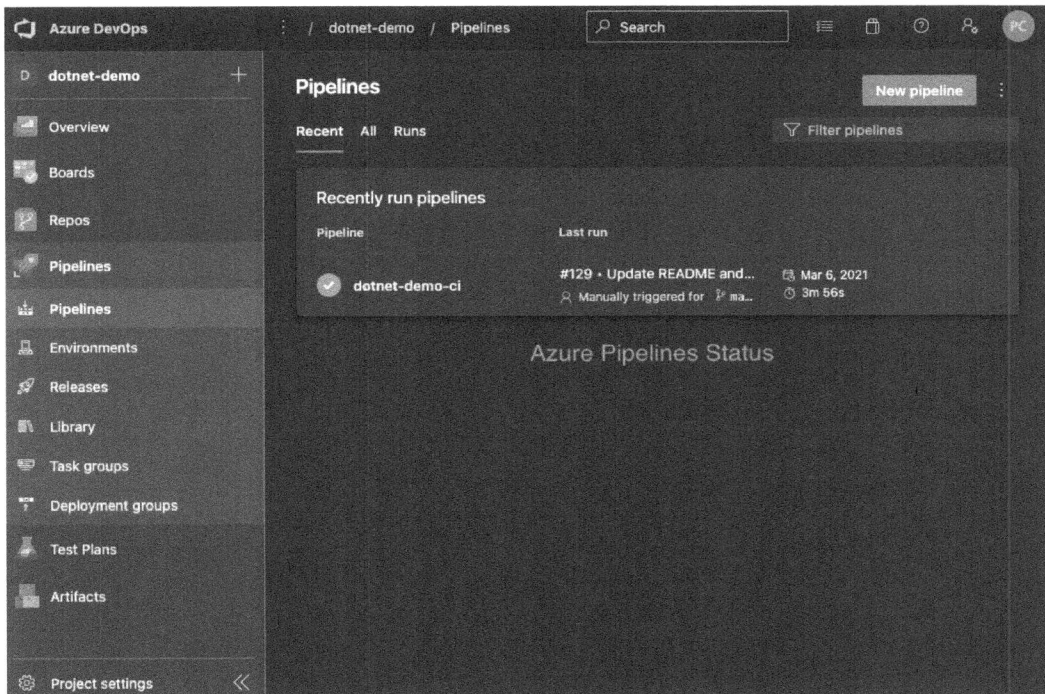

Figure 1.9 – The Azure pipelines' status

Clicking on the current pipeline status row will take you to the historical status of the pipeline. Two colors are used to indicate the status: green and red. These indicate successful and failed pipelines, respectively.

This section described all the components and their relationships. In the next section, you will understand the key differences between Azure Pipelines and other commonly used CI/CD tools.

Comparing Azure Pipelines with other CI/CD tools

Azure Pipelines has different features compared to other CI/CD services currently on the market. Let's take a closer look:

Features	Azure Pipelines	GitHub Actions	GitLab CI	Bitbucket Pipelines
`if` expression	X	X	-	-
Loop statement	X	-	-	-
Online service – CI/CD minutes free usage per month	1,800	2,000	400	50
Online service – free package storage (GB)	2	0.5	5	1
Self-hosted agents	X	X	X	X
Free users	5	Unlimited	Unlimited	5

Table 1.3 – CI/CD tools comparison

This will help you understand the important factors to consider when you're trying to decide on the right tool for your CI/CD platform.

Before you start to build a pipeline for application deployment, we must prepare the necessary agent pools, as demonstrated in the next section.

Setting up agent pools

Before using Azure Pipelines to build code and deploy code, you need at least one **build agent**. There are two build agent types: **Microsoft-hosted build agent**, which is included by default, and **self-hosted build agent**. Each agent type will be located under an **agent pool**, which is a collection of build and release agents.

A Microsoft-hosted build agent will be located under an agent pool called **Azure Pipelines**. You can create a new agent pool for self-hosted build agents and assign them under it.

To create pools, follow these steps:

1. Click on your project name in the web portal and click **Project settings | Agent pools | Add pool**:

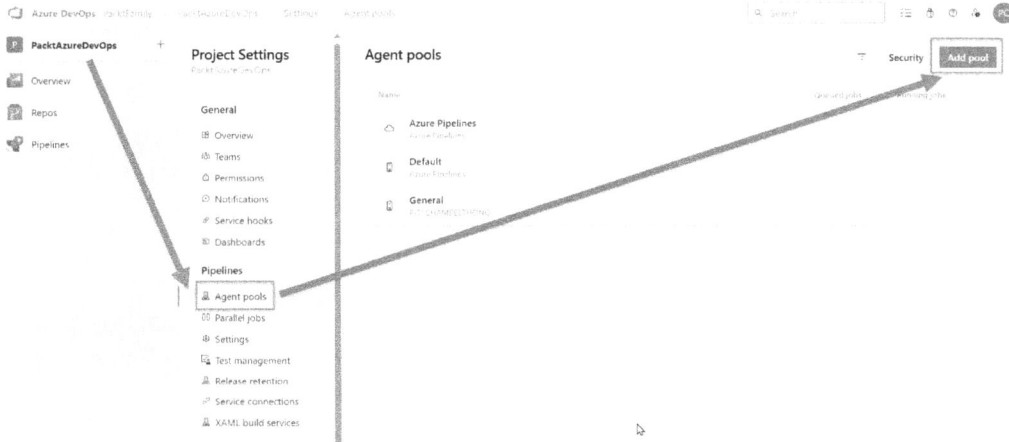

Figure 1.10 – Adding a pool

2. Enter the information shown in the following screenshot and then click **Create**:

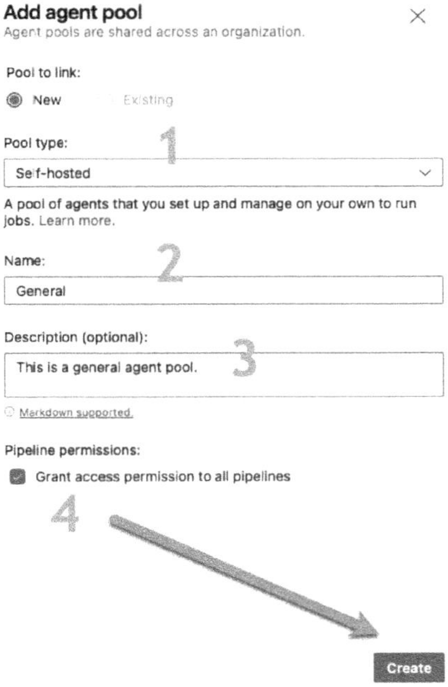

Figure 1.11 – Creating an agent pool

3.   Finally, you will see the new agent pool:

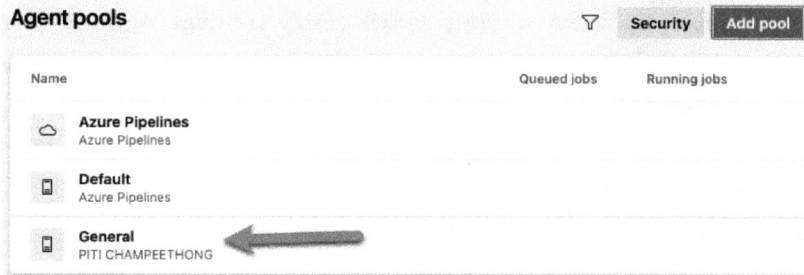

Figure 1.12 – Displaying the new agent pool

Once you've finished creating the new agent pool, you can start creating and setting up the self-hosted agent under a new agent pool. The following section will show you how to create a **personal access token (PAT)**.

Creating a PAT

Before you can create self-hosted agents on your server or machine, you must create a PAT. To do this, follow these instructions:

1.   Go to the **Settings** menu under your personal icon and click on **Personal access tokens**:

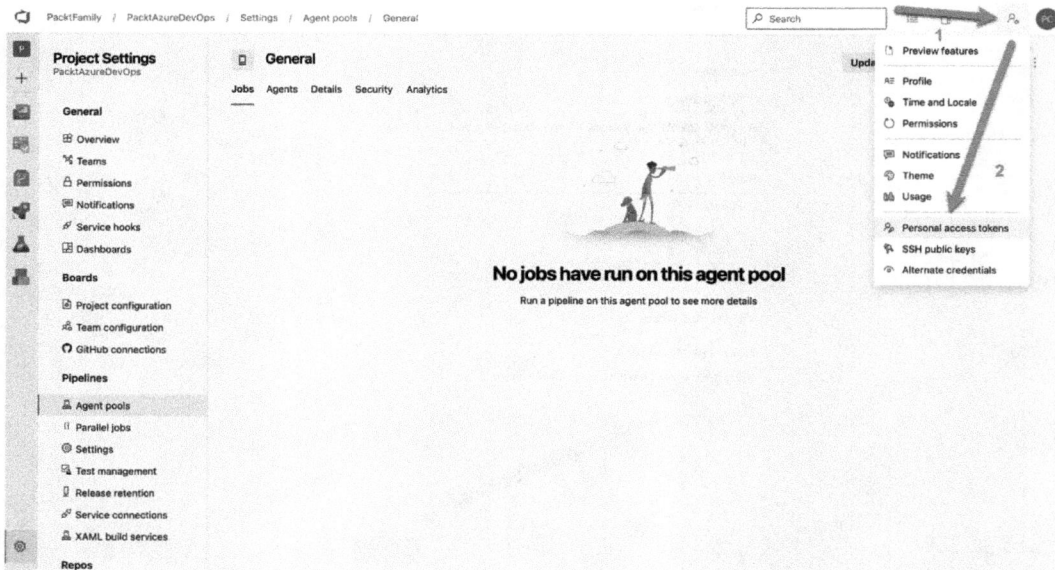

Figure 1.13 – Creating PATs

2. Click on **New Token**:

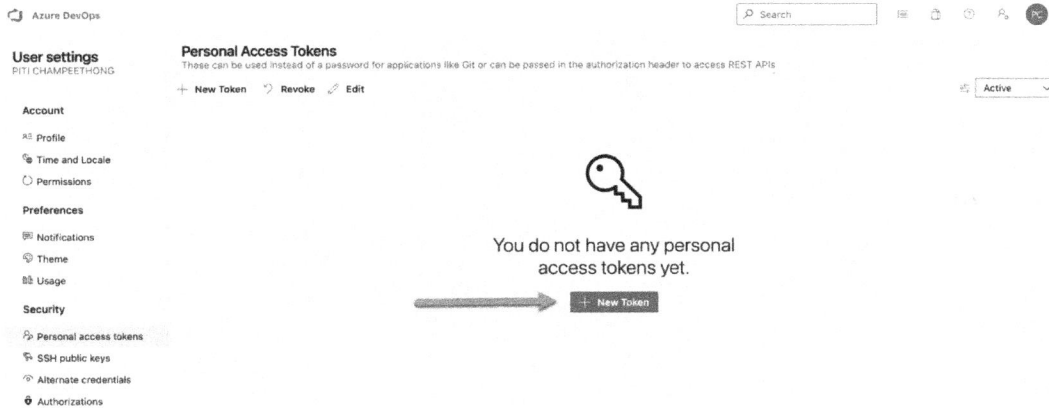

Figure 1.14 – New Token

3. Enter the required information:

    *   **Name**: Enter the name you need

    *   **Organization**: Select an organization you will link to

    *   **Expiration (UTC)**: There are four choices – 30 days, 60 days, 90 days, and custom-defined but under 2 years

    *   **Scopes**: Select **Custom defined | Agents Pools** with **Read & manage | Auditing** with **Read Audit Log**:

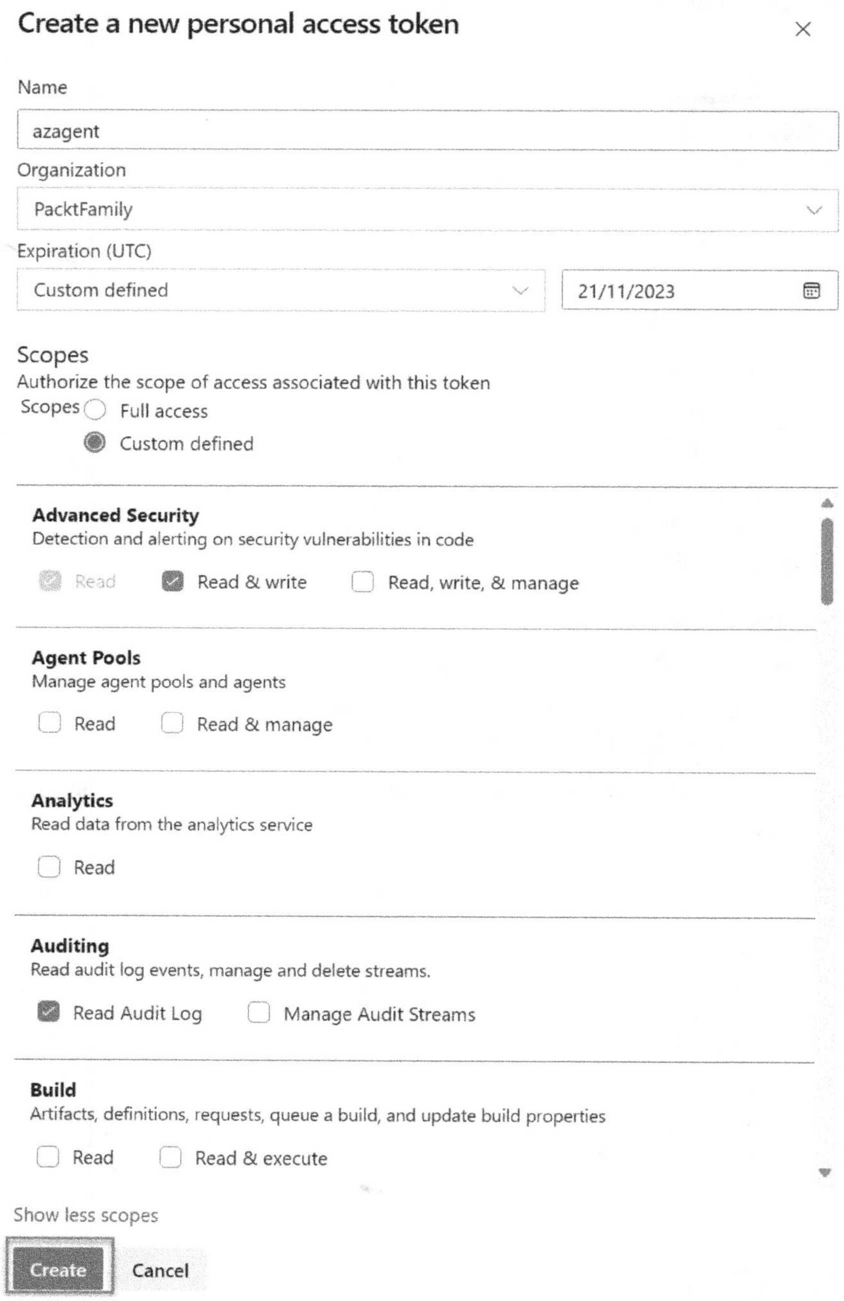

Create a new personal access token   ✕

Name

azagent

Organization

PacktFamily   ⌄

Expiration (UTC)

| Custom defined   ⌄ | 21/11/2023   🗓 |

Scopes
Authorize the scope of access associated with this token

Scopes ◯ Full access

◉ Custom defined

Advanced Security
Detection and alerting on security vulnerabilities in code

☑ Read   ☑ Read & write   ☐ Read, write, & manage

Agent Pools
Manage agent pools and agents

☐ Read   ☐ Read & manage

Analytics
Read data from the analytics service

☐ Read

Auditing
Read audit log events, manage and delete streams.

☑ Read Audit Log   ☐ Manage Audit Streams

Build
Artifacts, definitions, requests, queue a build, and update build properties

☐ Read   ☐ Read & execute

Show less scopes

[Create]   Cancel

Figure 1.15 – Entering the required information

4.   Copy the PAT before clicking on the **Close** button as you won't be able to see it again:

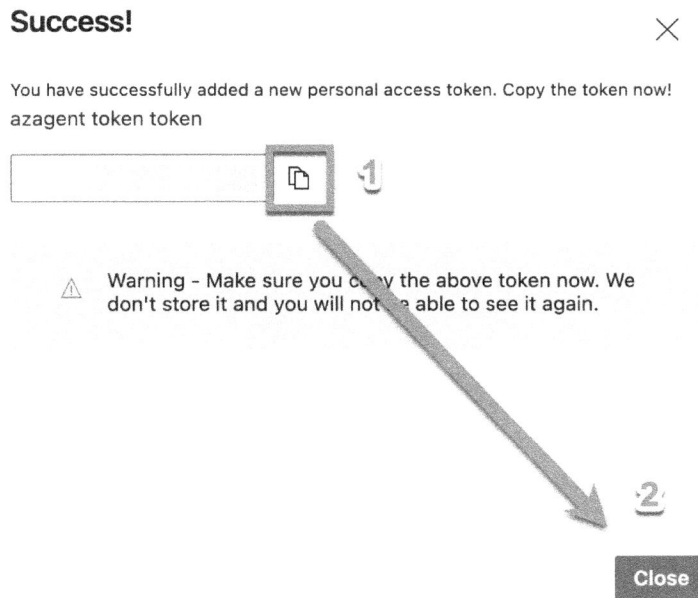

Figure 1.16 – Copying the PAT

Now, you are ready to set up a self-hosted agent.

Setting up self-hosted agents

After you've created a PAT, you can create a new self-hosted agent under a new agent pool. To do this, follow these steps:

1. Click on **PacktAzureDevOps | Agent pools | General**:

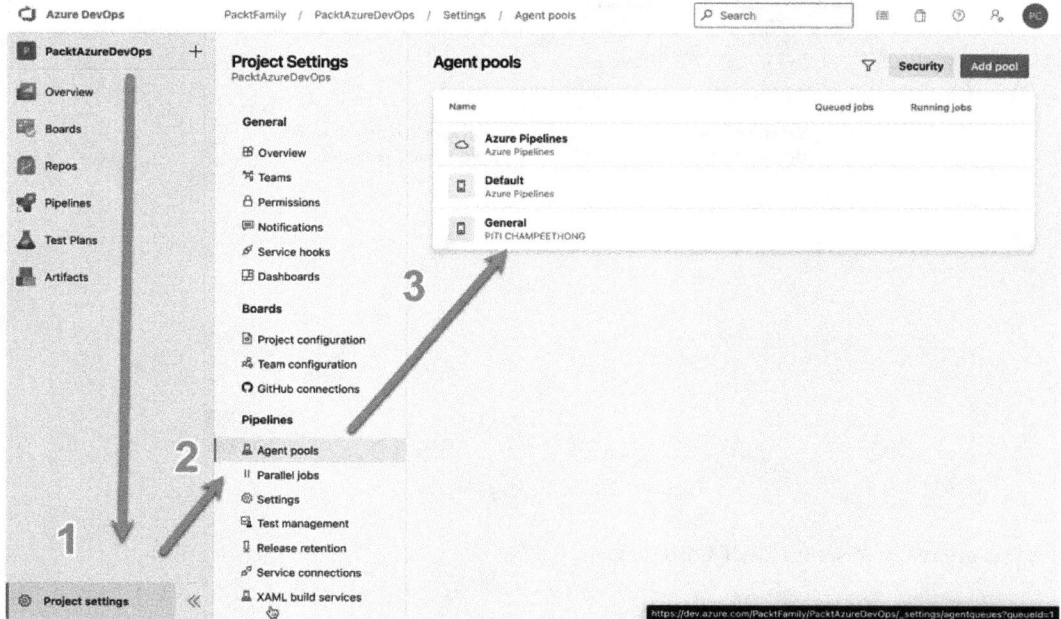

Figure 1.17 – Entering a new agent pool

2. Click on **New agent**:

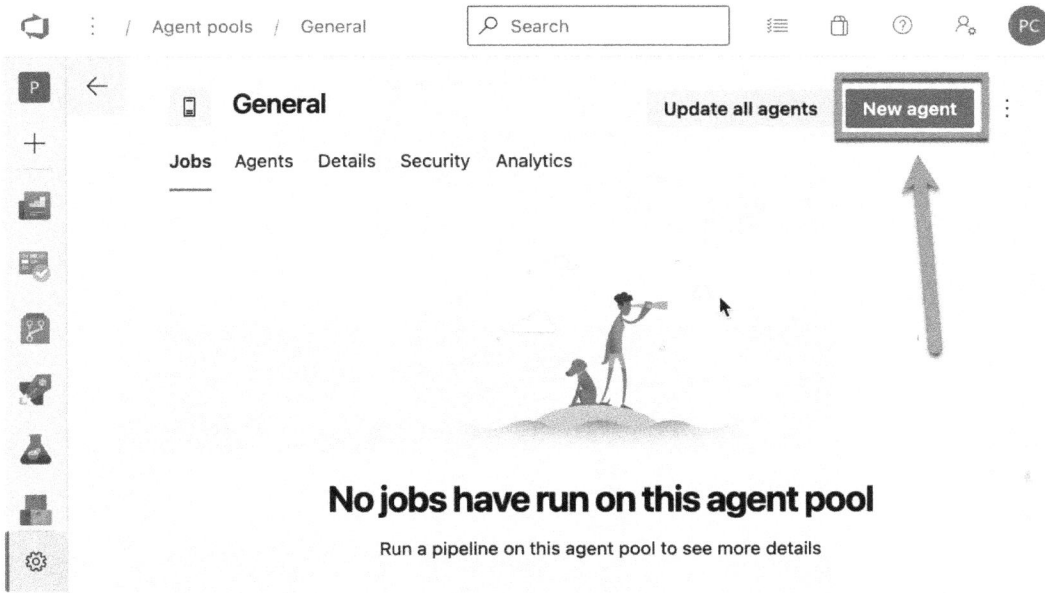

Figure 1.18 – Entering a new agent

3. You can download a self-hosted agent based on your operating system. Three operating system options will show you how to download and set them up:

  • **Windows users** can download build agent software from the **Windows** tab. There are two options, as shown in the following screenshot: Windows 64-bit (**x64**) and Windows 32-bit (**x86**):

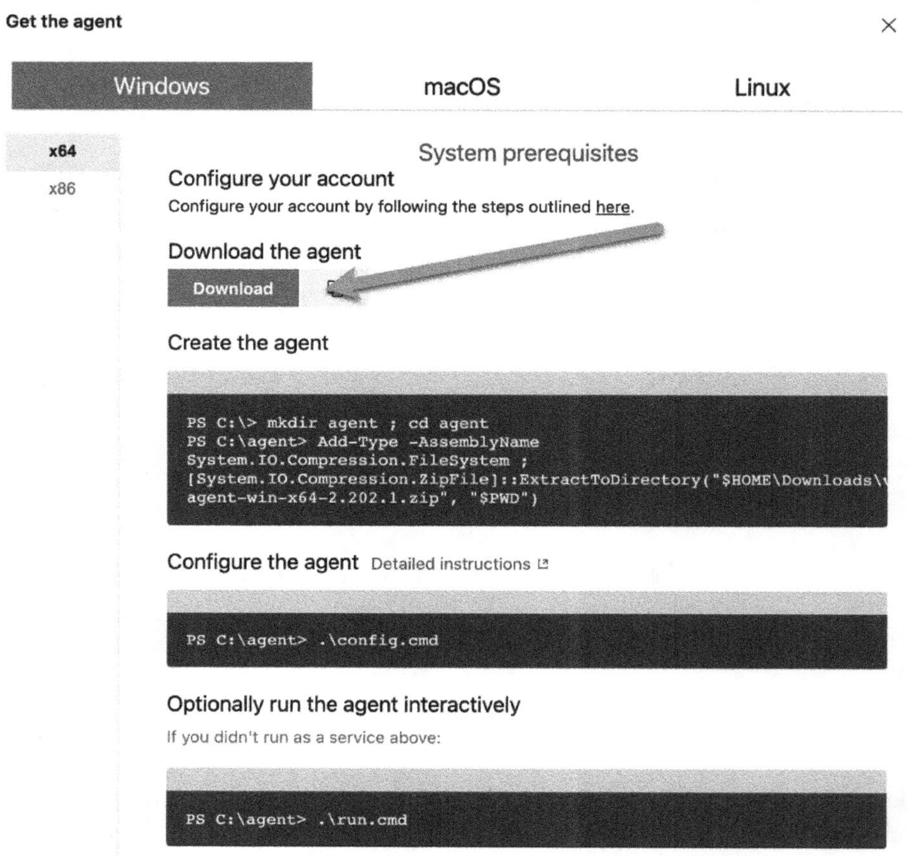

Get the agent   ✕

Windows	macOS	Linux

x64

x86

System prerequisites

Configure your account

Configure your account by following the steps outlined here.

Download the agent

Download

Create the agent

```
PS C:\> mkdir agent ; cd agent
PS C:\agent> Add-Type -AssemblyName
System.IO.Compression.FileSystem ;
[System.IO.Compression.ZipFile]::ExtractToDirectory("$HOME\Downloads\
agent-win-x64-2.202.1.zip", "$PWD")
```

Configure the agent Detailed instructions ⌤

```
PS C:\agent> .\config.cmd
```

Optionally run the agent interactively

If you didn't run as a service above:

```
PS C:\agent> .\run.cmd
```

Figure 1.19 – The Windows agent for setting up a file

- To set up a Windows agent, you need to run on PowerShell as an administrator.

- **Mac users** can download build agent software from the **macOS** tab:

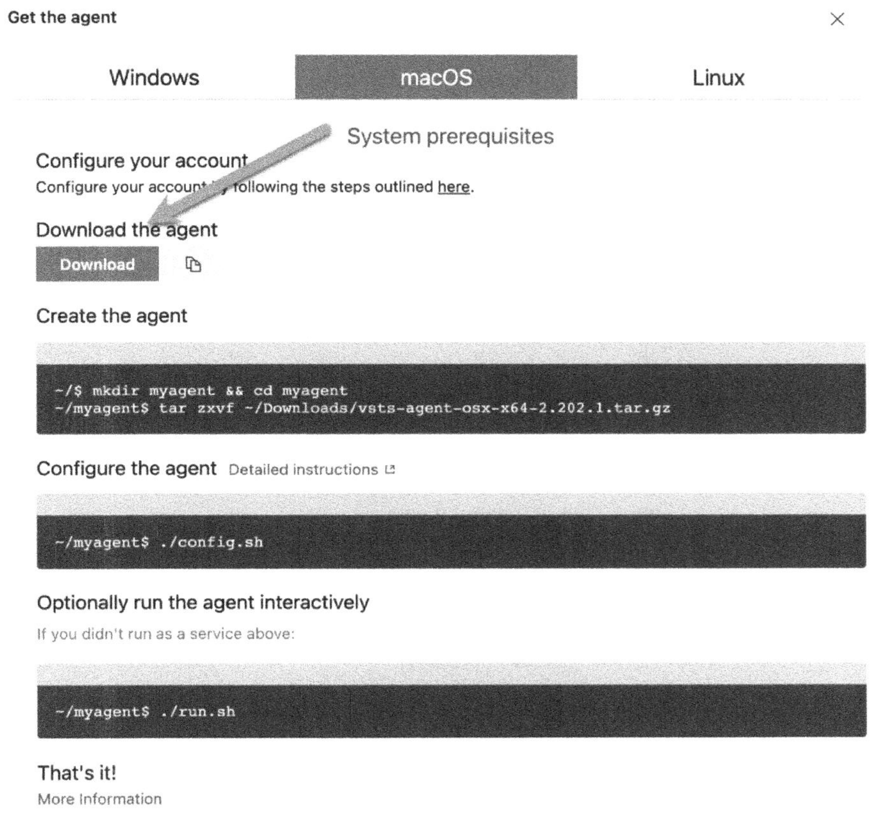

Figure 1.20 – macOS agent

- You don't need to use the `bash` command in the administrator role to set up the macOS agent.

- **Linux users** can download build agent software from the **Linux** tab. There are four options for computer architecture – **x64**, **ARM**, **ARM64**, and **RHEL6**:

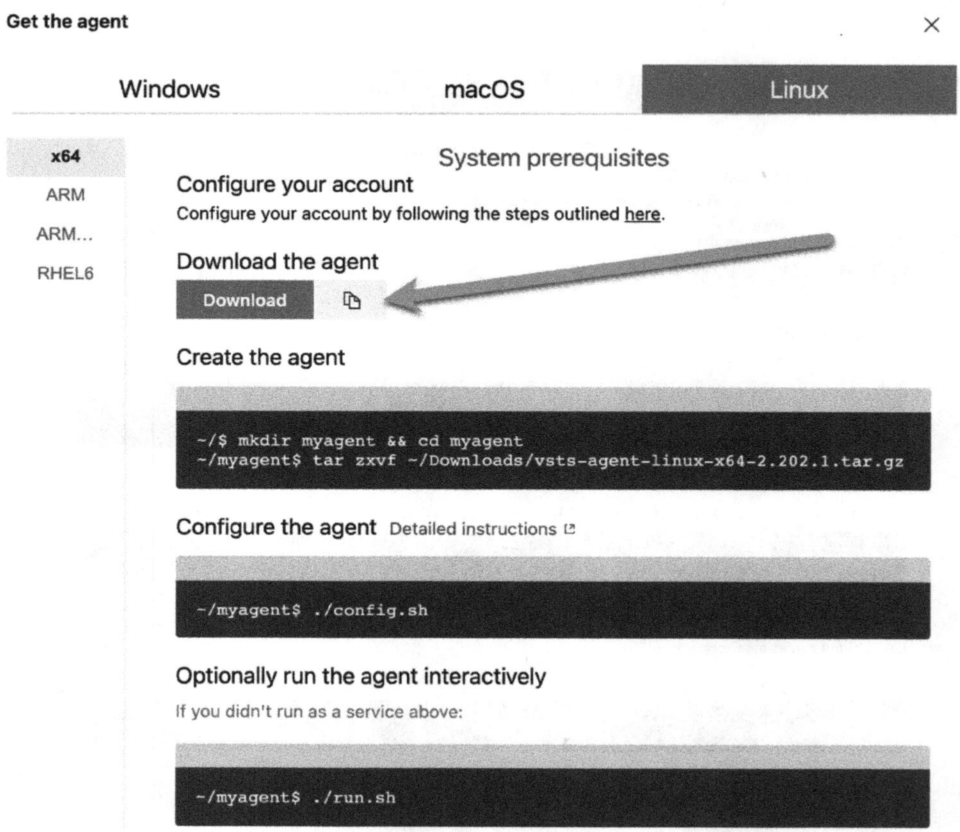

Get the agent   ×

| Windows | macOS | Linux |

x64
ARM
ARM...
RHEL6

System prerequisites

Configure your account
Configure your account by following the steps outlined here.

Download the agent

[Download]   ⎘

Create the agent

```
~/$ mkdir myagent && cd myagent
~/myagent$ tar zxvf ~/Downloads/vsts-agent-linux-x64-2.202.1.tar.gz
```

Configure the agent Detailed instructions ⤢

```
~/myagent$ ./config.sh
```

Optionally run the agent interactively
If you didn't run as a service above:

```
~/myagent$ ./run.sh
```

Figure 1.21 – Linux agent

- You don't need to use a `root` user for installation to set up a Linux agent.

4. After configuring the agent in each operating system, you must enter the following information:

```
Enter (Y/N) Accept the Team Explorer Everywhere license
agreement now? (press enter for N) > Y
Enter server URL > https://dev.azure.com/yourOrganization
Enter authentication type (press enter for PAT) > [ENTER]
Enter personal access token > [Personal Access Token]
Enter agent pool (press enter for default) > General
Enter agent name (press enter for [computer name]) > agent01
Enter work folder (press enter for _work) > [ENTER]
```

5. Once you start the service, you will see that `agent01` is active:

Figure 1.22 – The agent status dashboard

6. You can see the **Online** status of the build agent that has already been created:

Figure 1.23 – The action menu of the agent

7. You can delete the agent and **u**pdate to a new version of the agent by clicking on the button with the ellipses or three dots.

You are now ready to create the build and deployment on the `agent01` build agent. However, you need to set up the deployment group so that you can deploy your application on a local web server such as Microsoft **Internet and Information Services** (**IIS**). We'll do this in the next section.

Setting up deployment groups

Deployment groups are logical agent groupings for application deployment on target machines based on environments – they are typically named based on your project needs and promotion levels before they go to production. Agents are installed in each environment. Each agent under the deployment groups supports only Windows and Linux.

These deployment groups can be divided based on environments, such as development (*Dev*), quality assurance (*QA*), user acceptance testing (*UAT*), and production (*Prod*), as shown in the following figure:

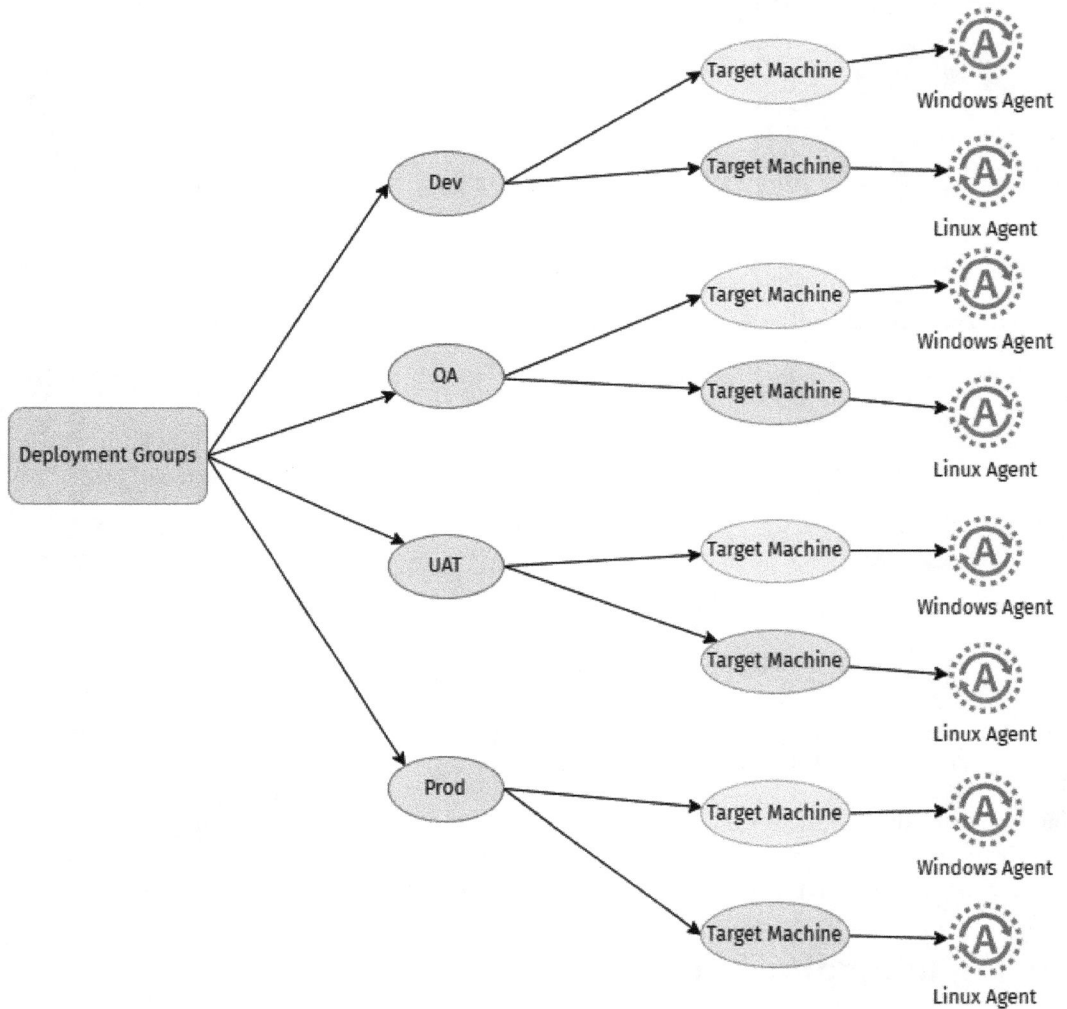

Figure 1.24 – Deployment groups concept

To create a deployment group, follow these instructions:

1. Navigate to **Pipelines** | **Deployment groups** | **Groups** | **Add a deployment group**:

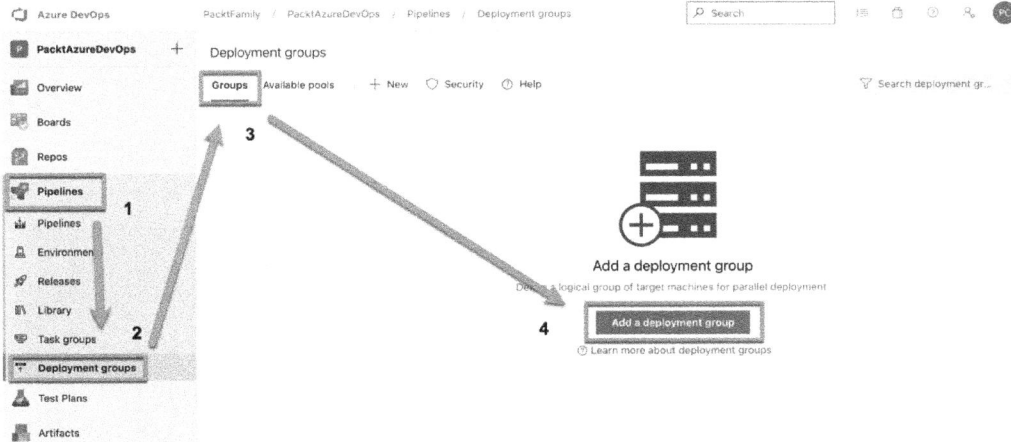

Figure 1.25 – Adding a deployment group

2. Enter the required information and click **Create**:

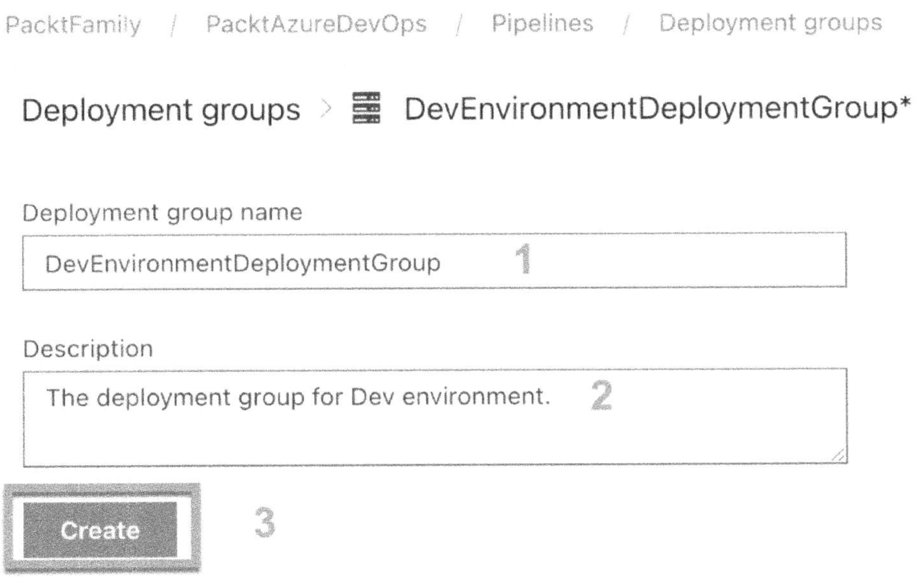

Figure 1.26 – Entering deployment group information

3. After creating the deployment group, you must set up the deployment agent under your deployment group. There are two operating systems the deployment agents support:

- **Windows users** can install an agent of the deployment group by copying the PowerShell script and running it with an administrator command prompt:

Figure 1.27 – Script to deploy an agent on Windows

- **Linux users** can install an agent of the deployment group by copying the bash shell script and running it with an administrator command prompt:

Figure 1.28 – Script to deploy an agent on Linux

4.   After setting up the agent for deployment groups, you will see that the build agent of the deployment group is online:

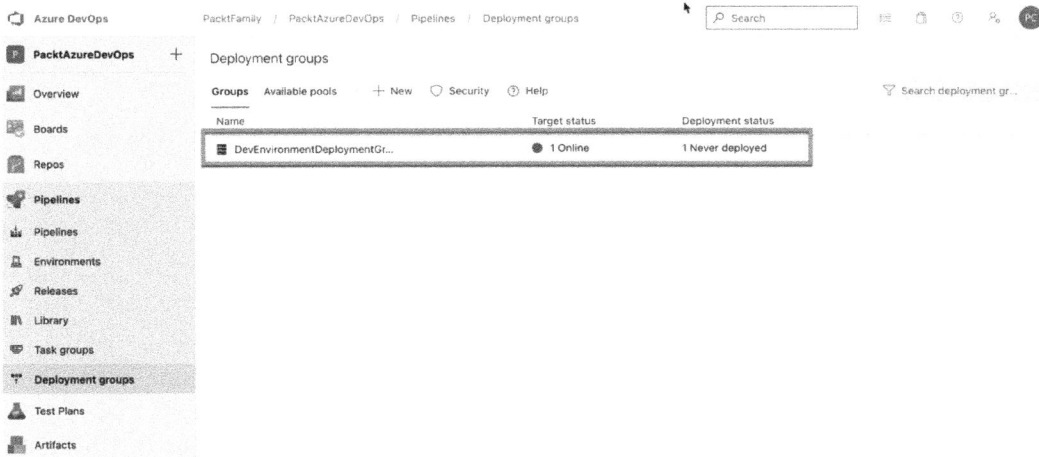

Figure 1.29 – The deployment agent is online

After setting up the build and deployment agent on a self-hosted computer, you are ready to create your first Azure pipeline.

Summary

In this chapter, you learned about the key concepts involved in CI/CD, which is a part of the Azure DevOps service from Microsoft. You learned about the basic structure of Azure Pipelines YAML and the difference between Azure Pipelines and other services currently on the market. You also learned about the fundamentals of Azure Pipelines, which will help you prepare your CI/CD pipelines for all real-world scenarios discussed later in this book. Finally, you learned how to set up agent pools, deployment groups, and self-hosted agents to prepare for the hands-on projects in the following chapters.

In the next chapter, you will apply the knowledge you've gained from this chapter to basic concepts and features and learn how to create a pipeline.

2

Creating Build Pipelines

In the previous chapter, we laid the groundwork for your Azure DevOps journey by preparing the build and deployment agents, pools, and deployment groups and explaining how to use them. This chapter will take the next step in helping you understand how to create your first pipeline. We will explore the fundamentals of using jobs, tasks, triggers, and stages and how to create build pipelines based on these concepts. This will enable you to streamline your workflow, minimize errors, and enhance productivity. It will also help make collaboration more efficient, and you'll be better prepared to implement practices such as CI/CD.

In this chapter, we will cover the following topics:

- Creating a build pipeline with a single job
- Creating tasks
- Creating multiple jobs
- Creating triggers
- Creating stages

Creating a build pipeline with a single job

After preparing the build and deployment agent in the previous chapter, this section will explain how to create a first **build pipeline** on the **Azure DevOps** portal. Before creating a first build pipeline, you must create an Azure repo for the source code repository. There are two options to create a build pipeline – the **classic editor**, which is the GUI editor for dragging and dropping and dropping the components to build a pipeline, and **Yet Another Markup Language** (**YAML**), which customizes the advanced Azure pipeline by the markup language. In this chapter, we will focus on using the classic editor.

Let's create an empty job to see how it works:

1.   Click on your project name from the web portal, and then click **Pipelines** | **Create Pipeline**:

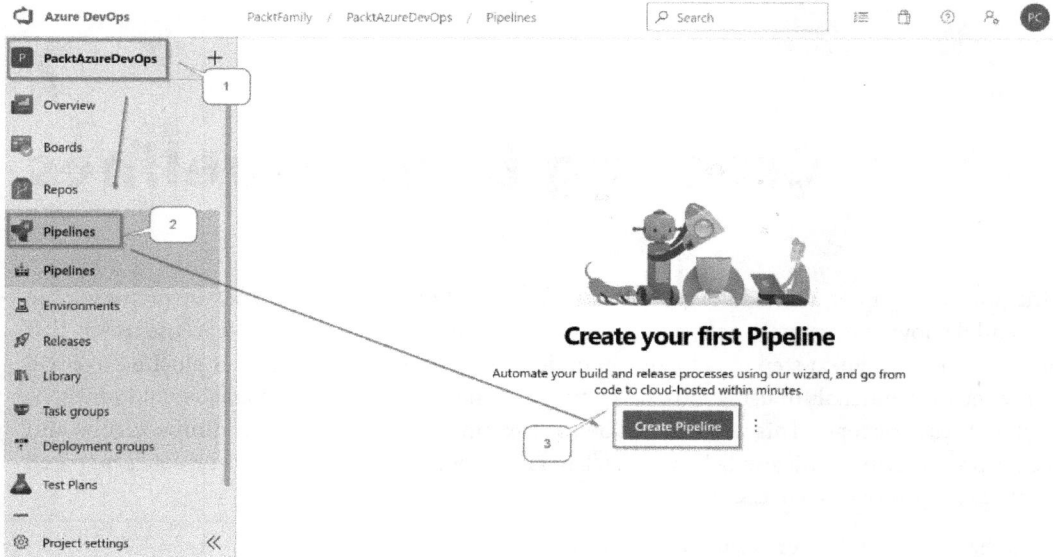

Figure 2.1 – Creating a new pipeline

2.   Click on **Use the classic editor**:

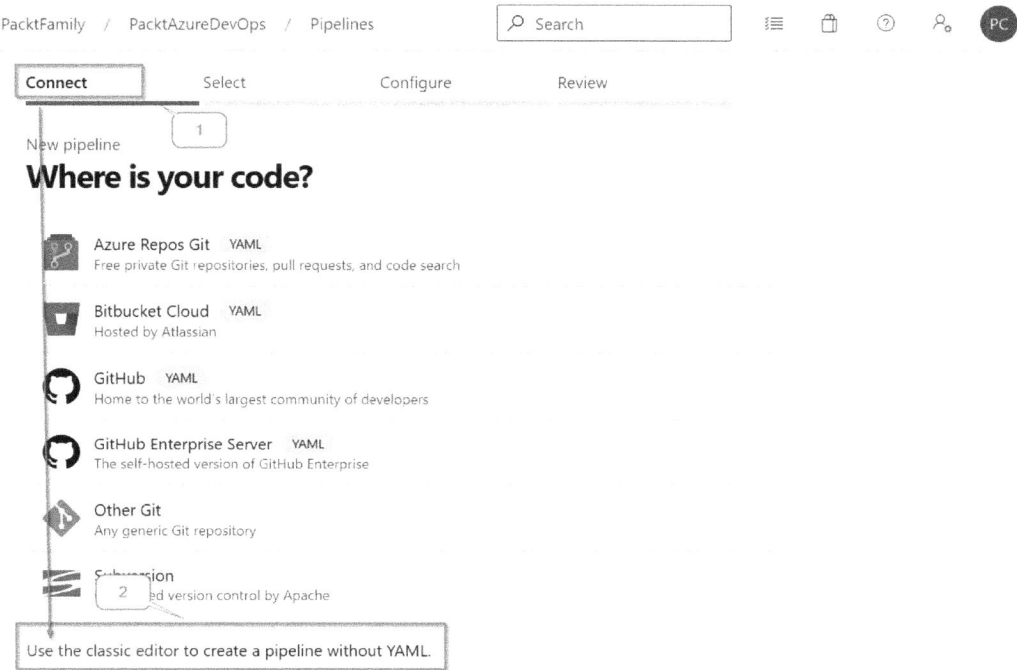

Figure 2.2 – Using the classic editor

3.   Select **Azure Repos Git**, a source code repository that is an existing service under Azure DevOps. This is quite compatible with other Azure DevOps services, such as Azure Pipelines, that are used in this demo. Select the options shown in the following screenshot for **Team project**, **Repository**, and **Default branch for manual and scheduled builds** to initiate an Azure pipeline, and then click on **Continue**:

Select a source

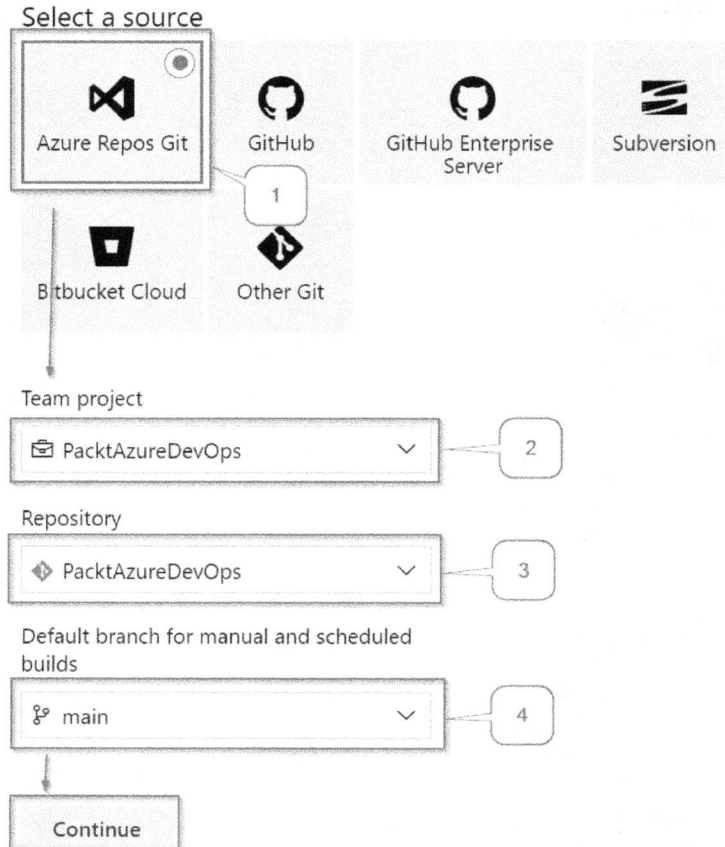

Figure 2.3 – Choosing the default branch

4. Click on **Empty Job**. The template provides a single job that contains tasks for your pipeline. We will start with an empty job so that you can learn the basic settings from the menu before we select all the templates that will be suitable for each project.

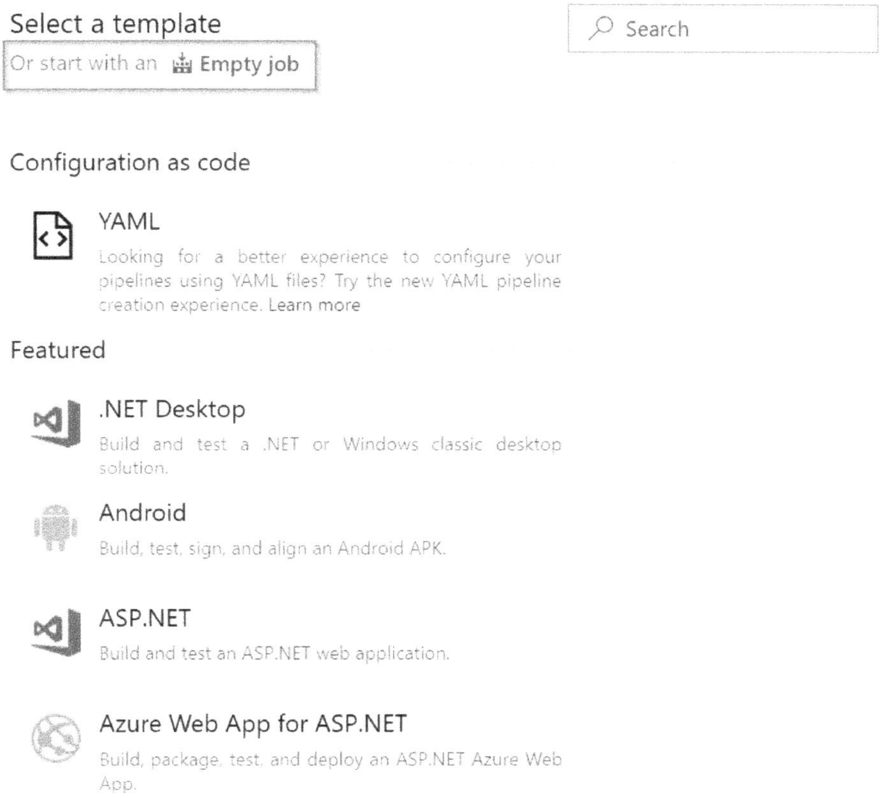

Select a template

Or start with an 🏗 **Empty job**

Configuration as code

YAML

Looking for a better experience to configure your pipelines using YAML files? Try the new YAML pipeline creation experience. Learn more

Featured

.NET Desktop

Build and test a .NET or Windows classic desktop solution.

Android

Build, test, sign, and align an Android APK.

ASP.NET

Build and test an ASP.NET web application.

Azure Web App for ASP.NET

Build, package, test, and deploy an ASP.NET Azure Web App.

Figure 2.4 – Selecting the build template

The other options shown in the screenshot include the following:

- **.NET Desktop**: The template for building pipelines to build and test .NET desktop solutions

- **Android**: The template for building pipelines to build and test Android APK files for Android applications

- **ASP.NET**: The template for building pipelines to build and test ASP.NET web applications

- **Azure Web App for ASP.Net**: The template for building pipelines to build, test, and deploy ASP.NET to Azure Web App services

5.   Click on the drop-down menu for **Save & queue**:

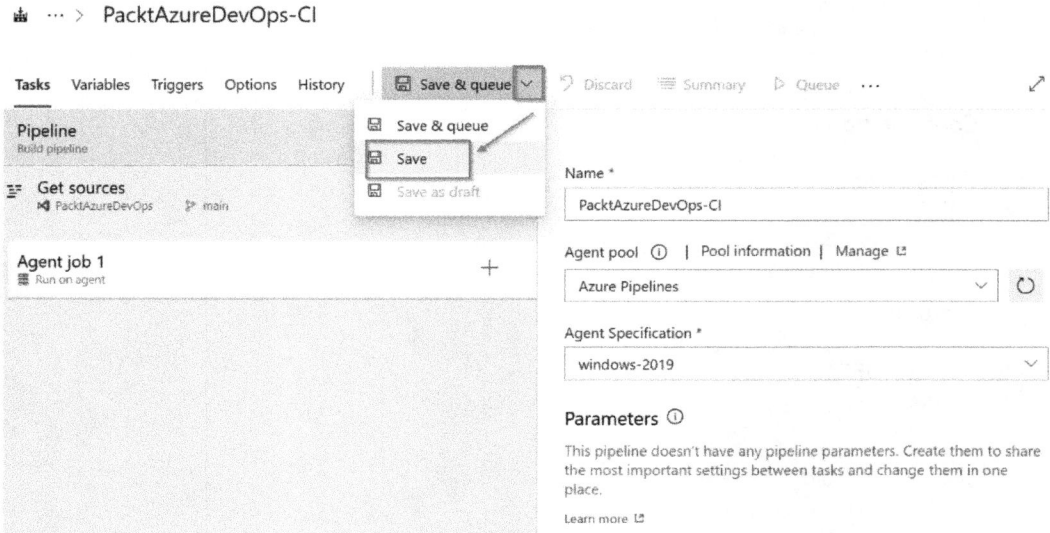

Figure 2.5 – Saving a build pipeline

6.   On the following screen, you can select a folder to save to and add a comment. Click on **Save**:

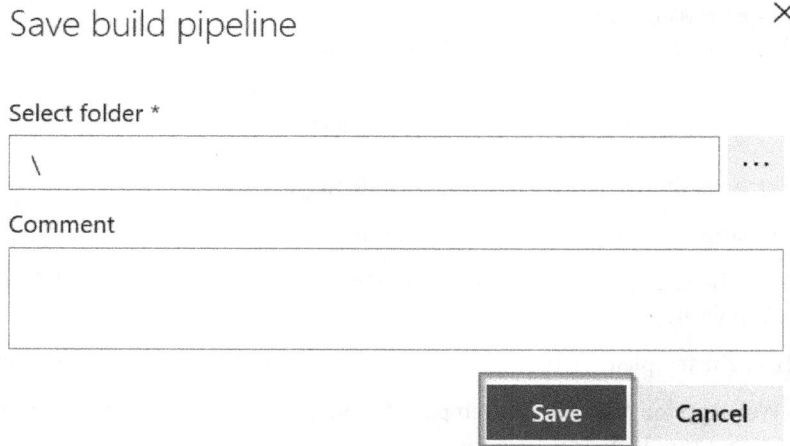

Figure 2.6 – Confirming a build pipeline

7. After clicking **Save**, a list of all the build pipelines will appear, as follows:

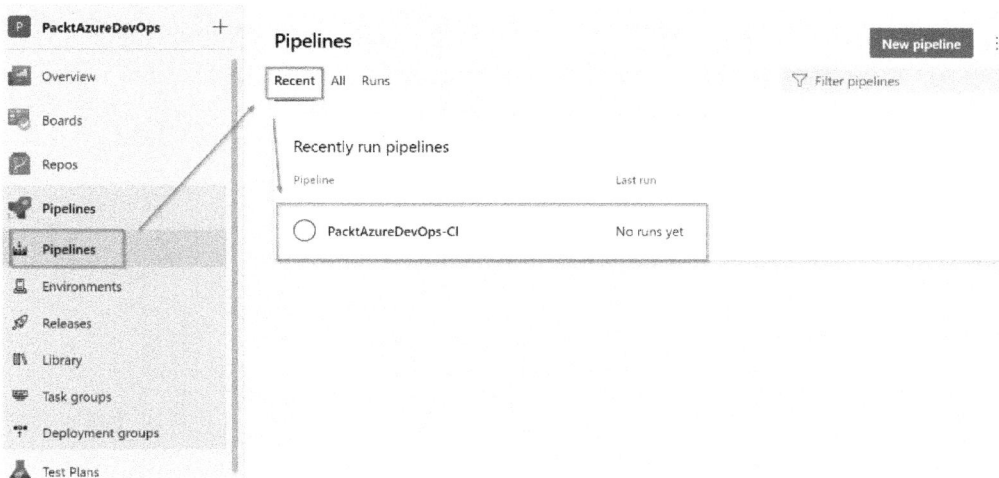

Figure 2.7 – The Pipelines dashboard

8. After you create a simple build pipeline, you can test it by clicking on the pipeline you want and clicking **Run pipeline**:

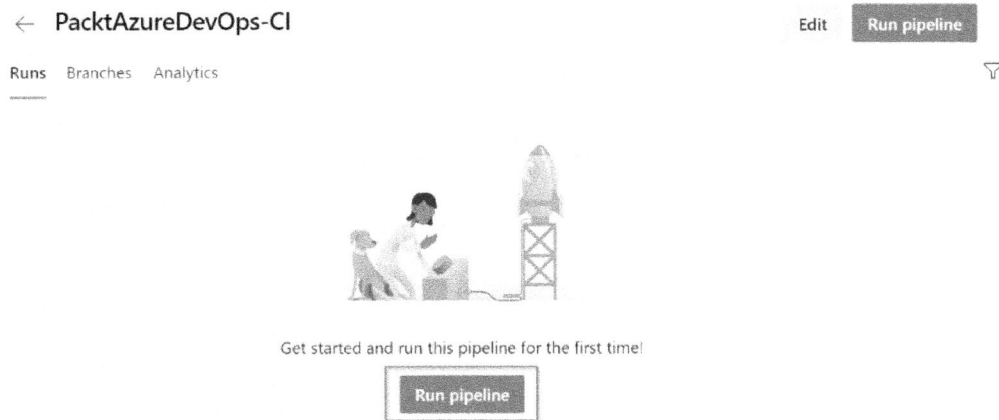

Figure 2.8 – Running the pipeline

9. You can set specific options before you run a build pipeline:

- **Agent pool**, where you can choose a Microsoft-hosted agent or a self-hosted agent
- **Agent Specification**, where you can choose an agent OS to run a build pipeline
- **Branch/tag**, where you can choose a build pipeline to run

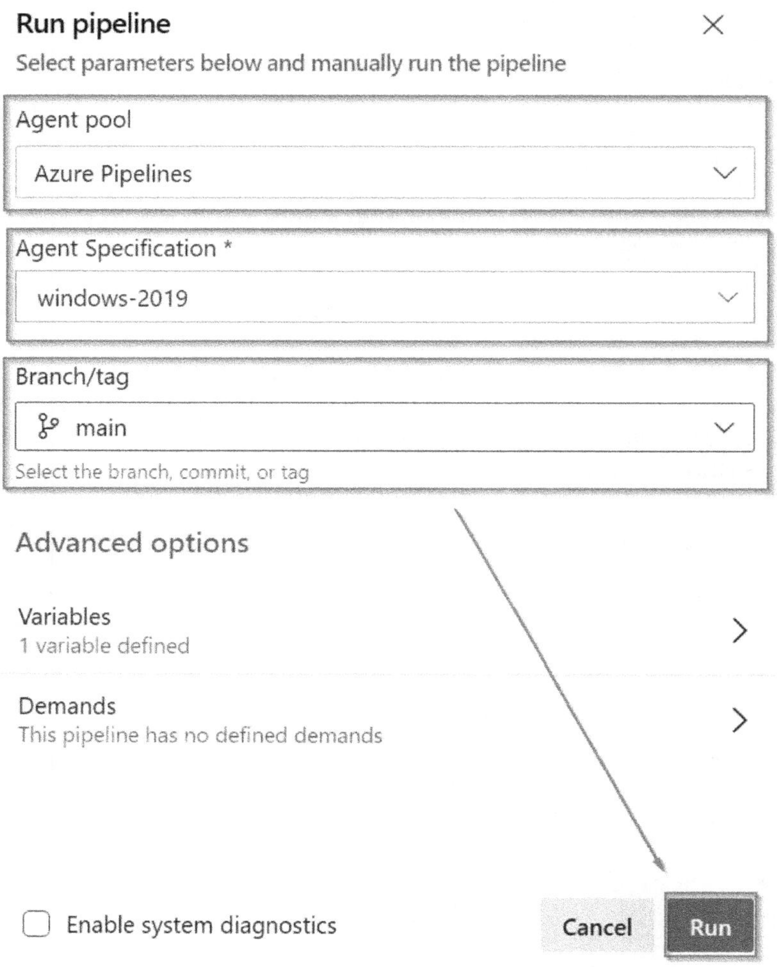

Figure 2.9 – The Run pipeline options

10. Finally, you can see the build result in detail and some summary information.

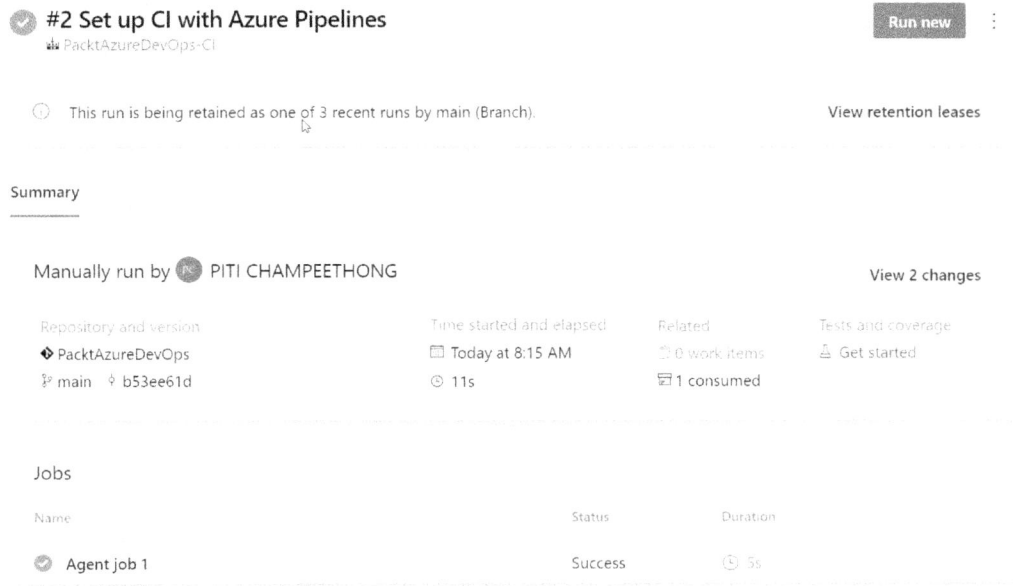

Figure 2.10 – A detailed build result

You can also see the latest status on the build pipeline row:

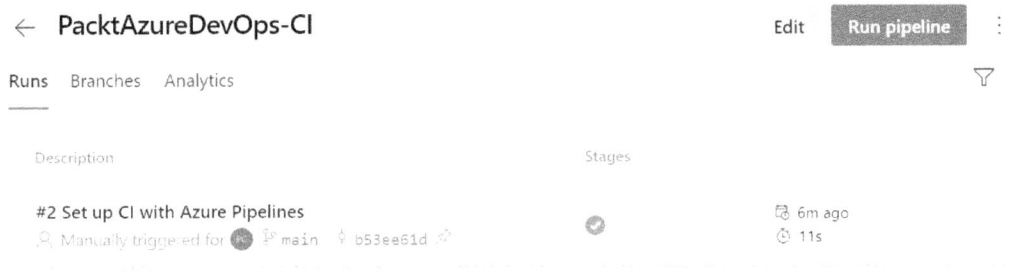

Figure 2.11 – A pipeline with the build result

In this section, you learned how to use the classic editor to make it easy to create a new build pipeline. In the next section, we will describe how to create tasks in this build pipeline.

Creating tasks

This section will teach you how to create **tasks** under a job. A build pipeline will contain one or more jobs, and each job will contain one or more tasks. In the previous section, you created a build pipeline that contains only one job. Follow these steps to create a task:

1. Edit a build pipeline by clicking on three dots symbols, and then on **Edit**.

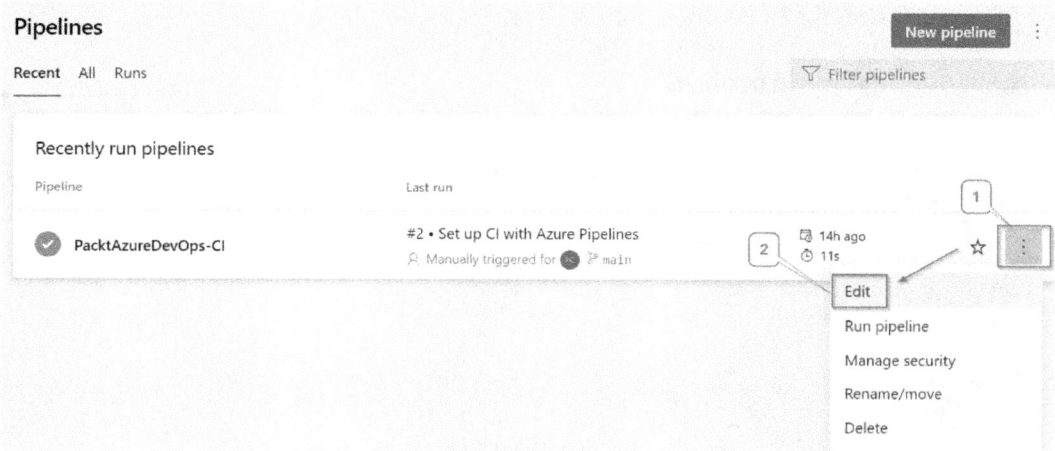

Figure 2.12 – Editing a build pipeline

2. After you edit a build pipeline, click on **Tasks** and then the + symbol, type `Command line` in the search box, and then click on **Add**:

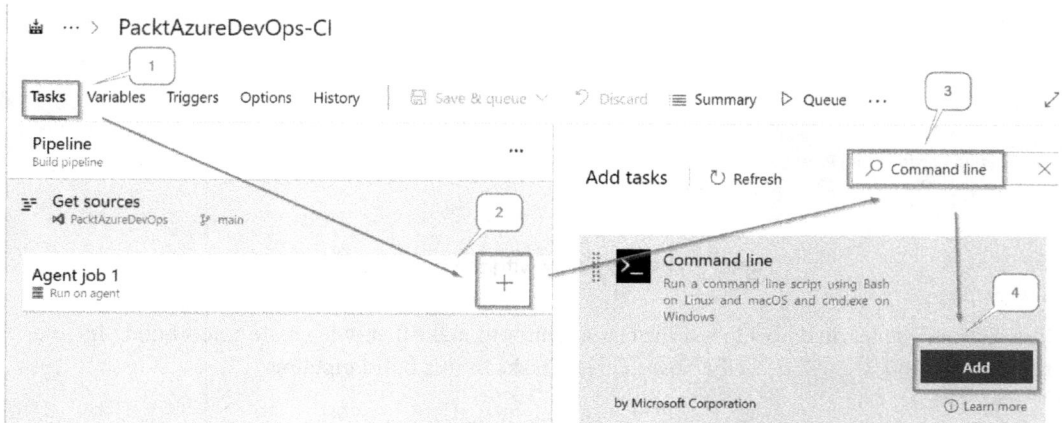

Figure 2.13 – Adding a command-line task

3. Click on the **Command line** task to enter the following details:

    - **Display name**: First Command Line Task

    - **Script**: For this demo, we will write a simple script to display the phrase *Hello First Task* –
      echo "Hello First Task":

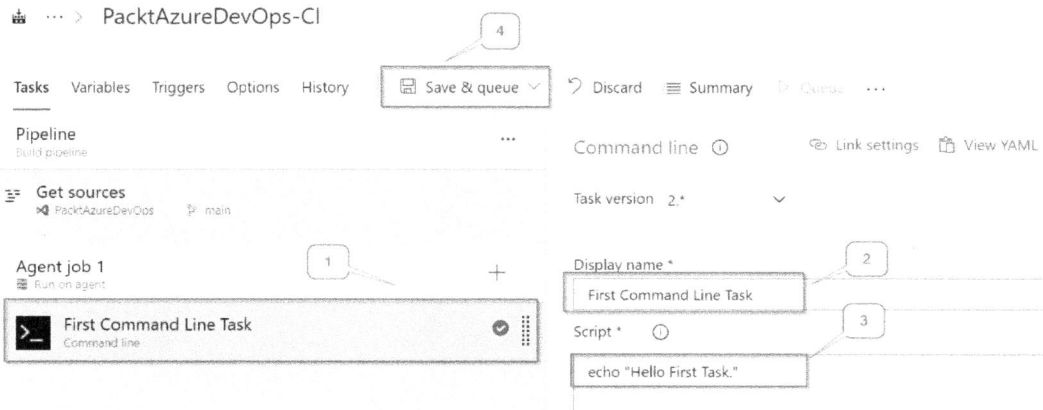

Figure 2.14 – Entering detail on the Command line task

Then, click **Save & queue**.

4. After you click on **Save & queue**, you will see the confirmation page. Then, click **Save and run**:

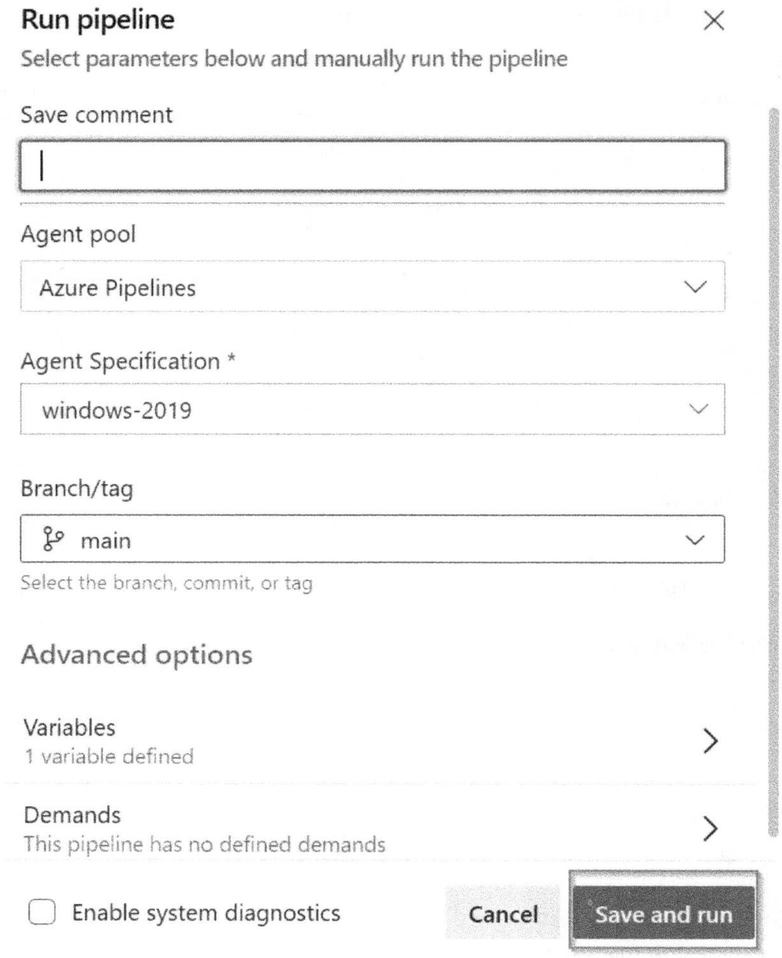

Run pipeline   ✕

Select parameters below and manually run the pipeline

Save comment

```
|
```

Agent pool

```
Azure Pipelines                                    ∨
```

Agent Specification *

```
windows-2019                                       ∨
```

Branch/tag

```
ꙮ  main                                            ∨
```

Select the branch, commit, or tag

Advanced options

Variables   >
1 variable defined

Demands   >
This pipeline has no defined demands

☐ Enable system diagnostics   Cancel   Save and run

Figure 2.15 – Saving and running a build pipeline

5. Before saving, however, you can also select **Advanced options**. Here, you can enter variables while you are building a pipeline instead of preparing them in advance. You can see the available options in the following screenshot:

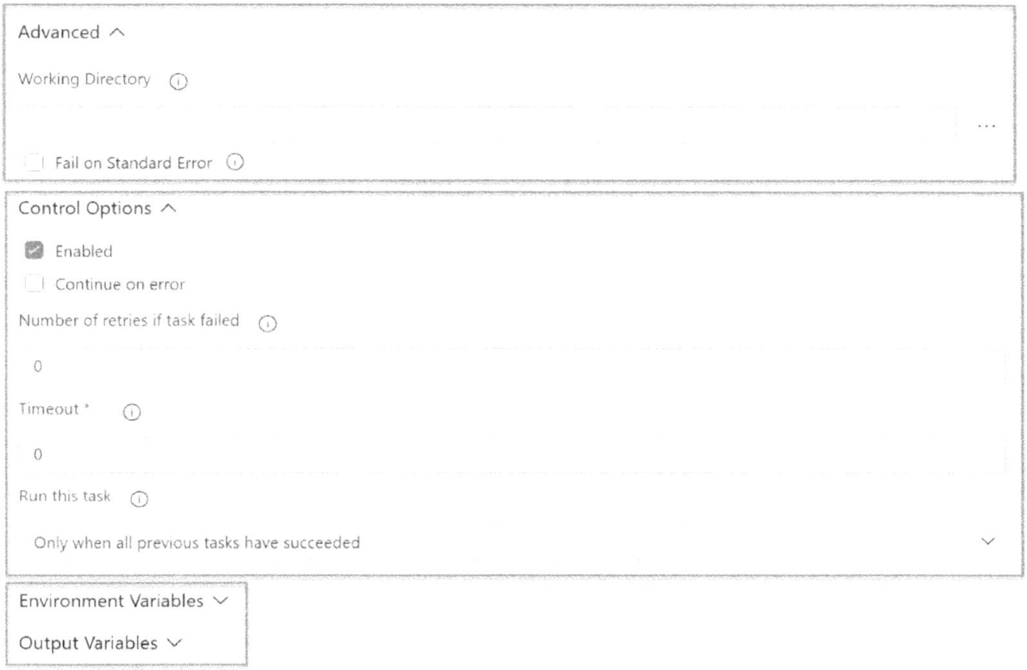

Figure 2.16 – Advanced configuration

Let's look at these options more closely:

- **Advanced**:

    - **Working Directory**: You can enter any path to run the command line task. If you don't enter any path, it will run on the root path.

    - **Fail on Standard Error**: Turn on this feature when you need to stop a task if any error happens inside it. For example, if you run a command to read an email from the internet and the internet disconnects, this task will display errors.

- **Control Options**:

    - **Enabled**: Turn this on when you need to enable control options:

    - **Continue on error**: Turn on when you need to continue running the next task if it finds an error.

    - **Number of retries if task failed**: Enter the number of times you need to rerun a task after it fails.

    - **Timeout**: Enter the number of minutes this task will run for before it is canceled.

- **Run this task**:

  - **Only when all previous tasks have succeeded**

  - **Even if a previous task has failed, unless the build was canceled**

  - **Even if a previous task has failed, even if the build was canceled**

  - **Only when a previous task has failed**

  - **Custom conditions**: This is where you can create a rule if the aforementioned options don't match your needs.

- **Environment Variables**: You can add additional variables when you run the command. For example, you can add a URL when you run this command to download the file.

- **Output Variables**: You can assign the output from this task and carry it to the next task. For example, you can pass the token login in the output variable after this task to log in successfully to the next task.

6. Once you save these options, you can see the new history of a build pipeline:

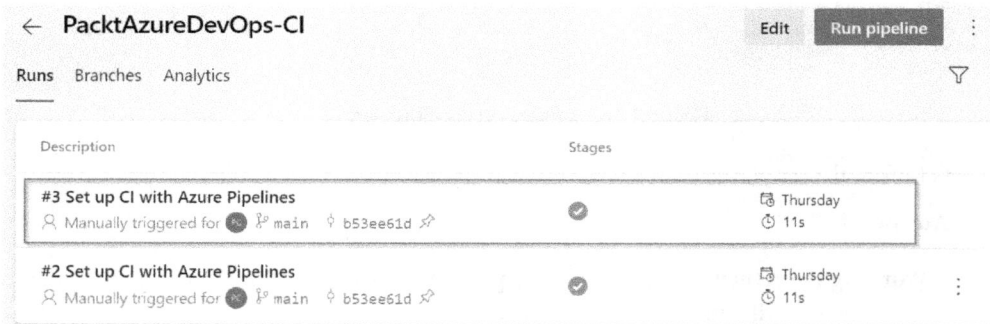

Figure 2.17 – Displaying the history of a build pipeline

7. Click on the row where you want to see the task details. Here, you will see the output of the running command for each task:

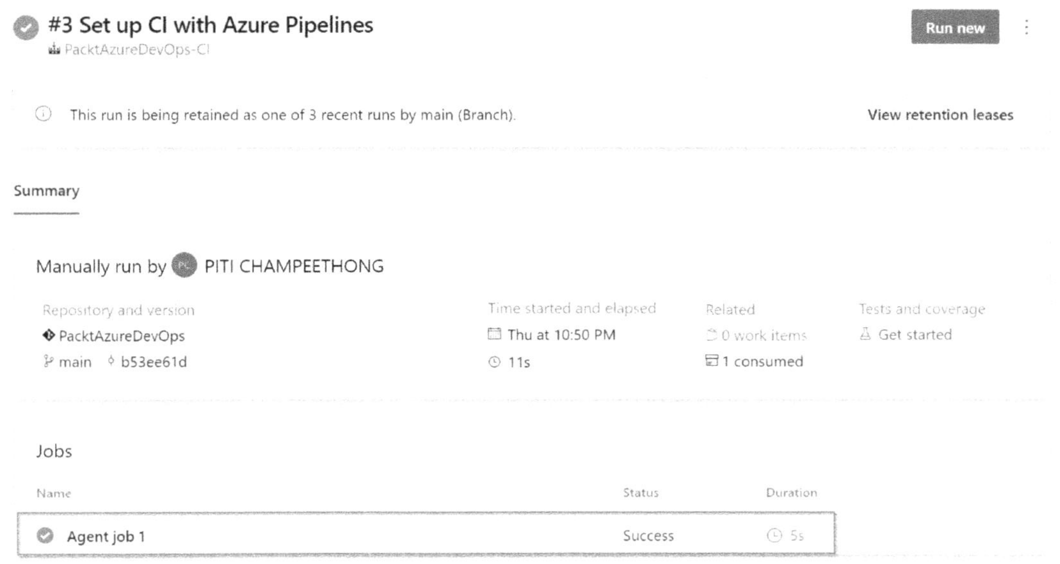
#3 Set up CI with Azure Pipelines
PacktAzureDevOps-CI
Run new
This run is being retained as one of 3 recent runs by main (Branch).
View retention leases
Summary
Manually run by PITI CHAMPEETHONG
Repository and version
PacktAzureDevOps
main b53ee61d
Time started and elapsed
Thu at 10:50 PM
11s
Related
0 work items
1 consumed
Tests and coverage
Get started
Jobs
Name
Status
Duration
Agent job 1
Success
5s

Figure 2.18 – The job status

8. Click on **Agent job 1** to see the job details:

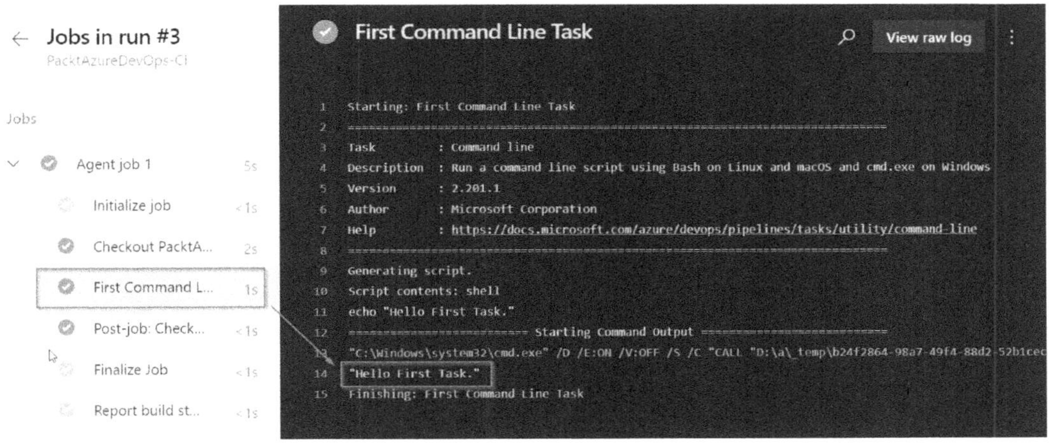
Jobs in run #3
PacktAzureDevOps-CI
Jobs
Agent job 1 5s
Initialize job <1s
Checkout PacktA... 2s
First Command L... 1s
Post-job: Check... <1s
Finalize Job <1s
Report build st... <1s
First Command Line Task
View raw log
Starting: First Command Line Task
Task : Command line
Description : Run a command line script using Bash on Linux and macOS and cmd.exe on Windows
Version : 2.201.1
Author : Microsoft Corporation
Help : https://docs.microsoft.com/azure/devops/pipelines/tasks/utility/command-line
Generating script.
Script contents: shell
echo "Hello First Task."
Starting Command Output
"Hello First Task."
Finishing: First Command Line Task

Figure 2.19 – Job details with tasks

In this section, you learned how to create a task under a build pipeline and view the job status and task details. This can help you ensure that all the tasks are as expected. The next section will teach you how to create multiple jobs in a build pipeline.

Creating multiple jobs

Sometimes, when you create a build pipeline for an application that requires building its code for different operating systems simultaneously, you need to create another job for that purpose. There are two job types – **agent job** and **agentless job**. An agent job is a job that needs to run on an agent or target computer, and an agentless job will run on the Azure DevOps application server directly. The following steps describe how to create other agent jobs under a build pipeline:

1. Edit a build pipeline by clicking on the **...** symbol and then on **Edit**:

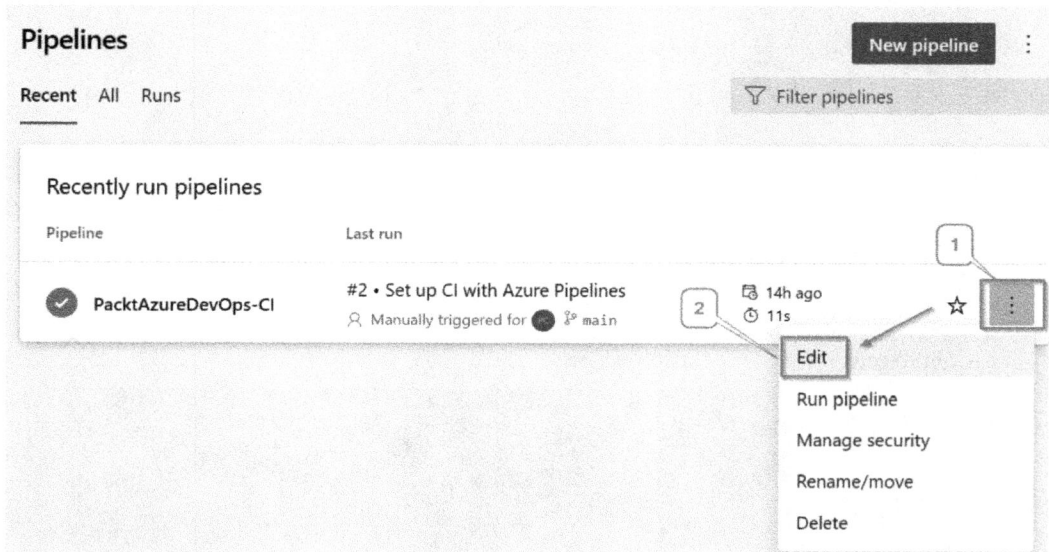

Figure 2.20 – Editing a build pipeline

2. Click on the **...** button and select **Add an agent job**:

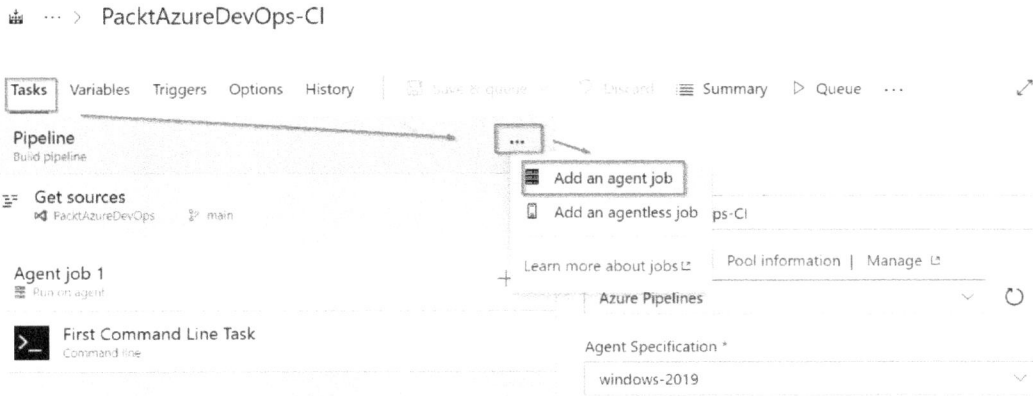

Figure 2.21 – Adding an agent job

3. Click on **Agent job** and update the job information as follows:

- **Display name**: Agent job 2
- **Agent pool**: **Azure Pipelines**
- **Agent Specification**: **ubuntu-latest**

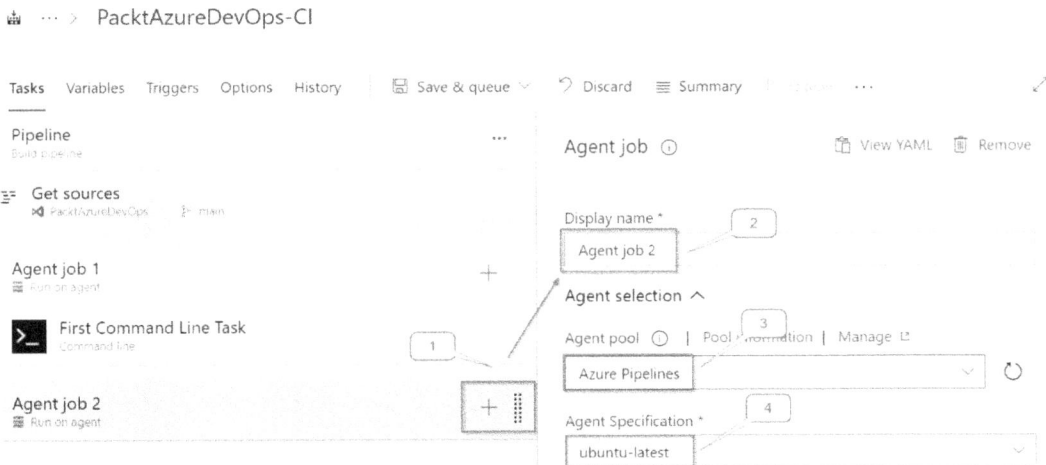

Figure 2.22 – Adding another job

4. Click on the + button for the job, type `Command line` in the search box, and then click the **Add** button:

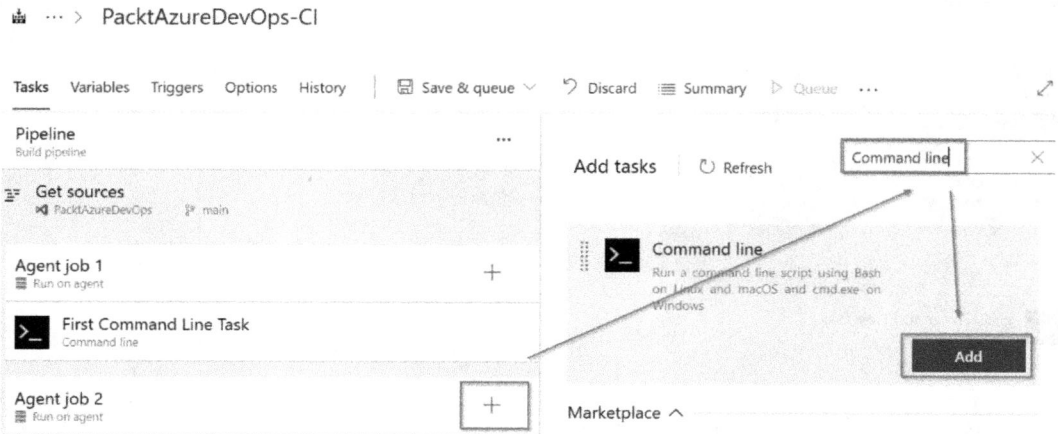

Figure 2.23 – Adding a new task

5. Click on the row with the **Command Line Task** and update the following fields:

    -   **Task version**: This is a package version of the Command Line Task. We usually use the latest version. In our case, this is version 2.

    -   **Display name**: Add a suitable name for your task. In our case, this is `Second Command Line Task`.

    -   **Script**: Add the following basic script – `echo "Hello Second Task on Linux"`. This will print out this text on the resultant page after you have finished running a pipeline:

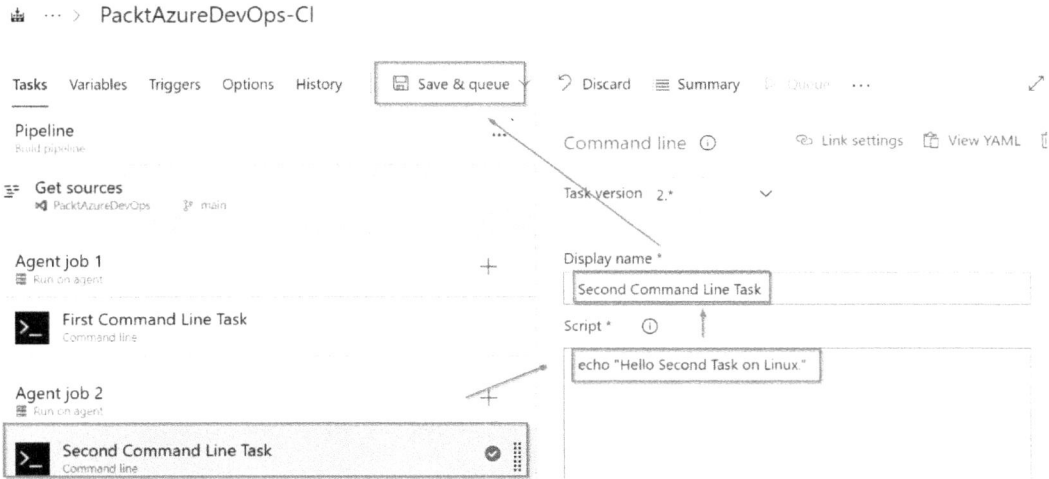

Figure 2.24 – Updating the second command line task

6.   After you click **Save & queue**, you will see the following result:

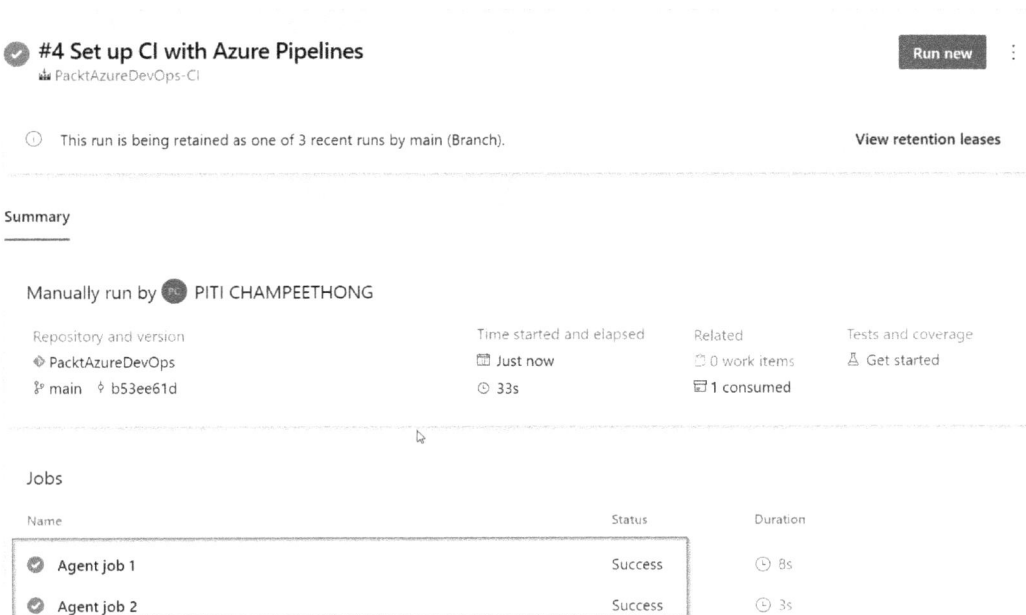

Figure 2.25 – The result of two agent jobs

> **Pro tip**
>
> For error handling, when you have more than one agent Job, you can set up parameters in advanced configuration to ensure that **Agent job 2** will run if **Agent job 1** is completed successfully.

Let's look at some of the advanced options available in the **Agent job** properties, as shown in the following screenshot:

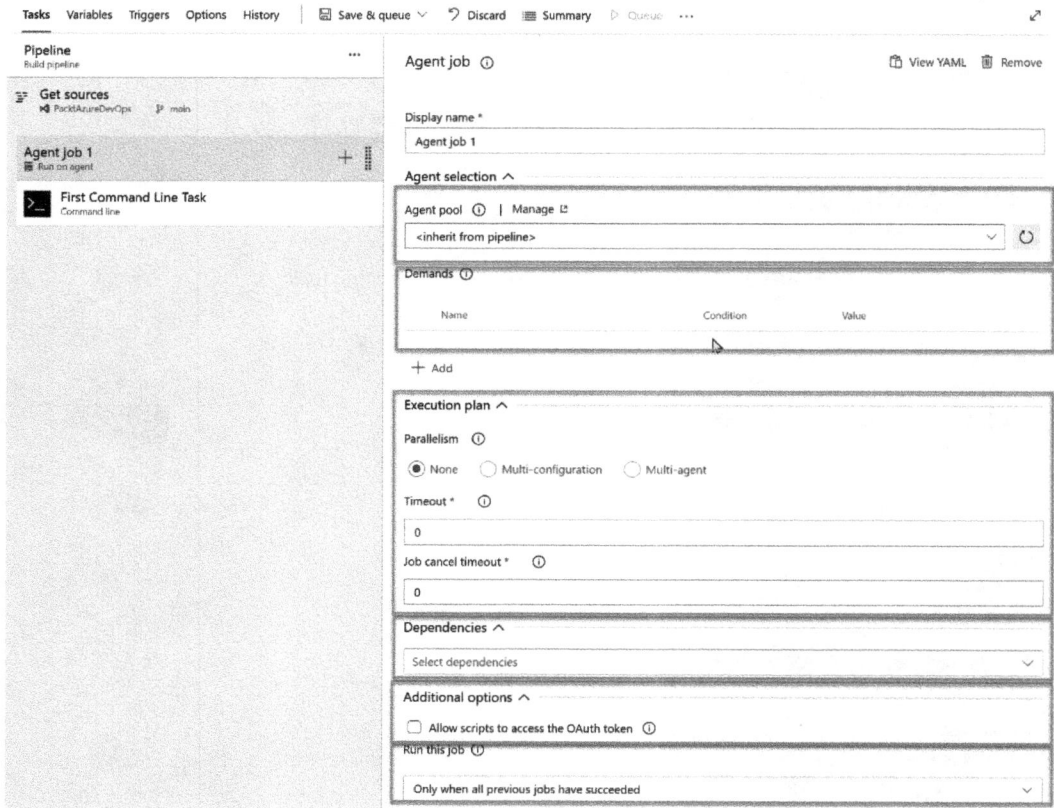

Figure 2.26 – Advanced options for an agent job

Let's look at these options in some detail:

- **Agent selection**:

  - **Agent pool**: Select a custom or default agent to build jobs.

  - **Demands**: Add condition parameters to allow only agents that meet conditions to run this agent job. For example, only a Linux agent can run this job.

- **Execution plan**:

  - **Parallelism**: The options available here are the following:

    - **None**: No jobs will run parallelly.

    - **Multi-configuration**: Turn this on when you have different configurations to run for each agent job. For example, you may need to run a job test on three browsers, so you will need Chrome, Edge, and Firefox browsers.

    - **Multi-agent**: Turn this on when you have many agents and need to use them to run an agent job.

  - **Timeout**: Enter a minute number; an agent job is allowed to execute on the agent before being canceled.

  - **Job cancel timeout**: Enter a number of minutes before the agent job is canceled after getting a cancel request from the agent.

- **Dependencies**: You can select a previous agent job when you need it completed successfully before running another agent job.

- **Additional options**:

  - **Allow scripts to access the OAuth token**: turn on when you need to use the OAuth token to pass to another agent job by using REST API.

- **Run this job**:

  - **Only when all previous jobs have succeeded**

  - **Even if a previous job has failed**

  - **Only when a previous job has failed**

  - **Custom condition using variable expressions**: For example, **succeeded()** means the agent job will run if the previous one completely successfully

In this section, you learned how to create a second job to separate the Linux operation system. Two jobs can run simultaneously, and this use case is suitable for running a task in different operating systems without depending on each other, such as when you create a build pipeline that needs to deploy your application to Google Play and Apple Store simultaneously without running in sequence.

The following section will teach you how to create a trigger to make a build pipeline run automatically.

Creating triggers

This section will teach you how to create a **trigger** – that is, make a build pipeline run automatically when you push your code to a specific branch:

1. Edit a build pipeline by clicking on the three dots to view the details of the build pipeline, and then click on **Edit**:

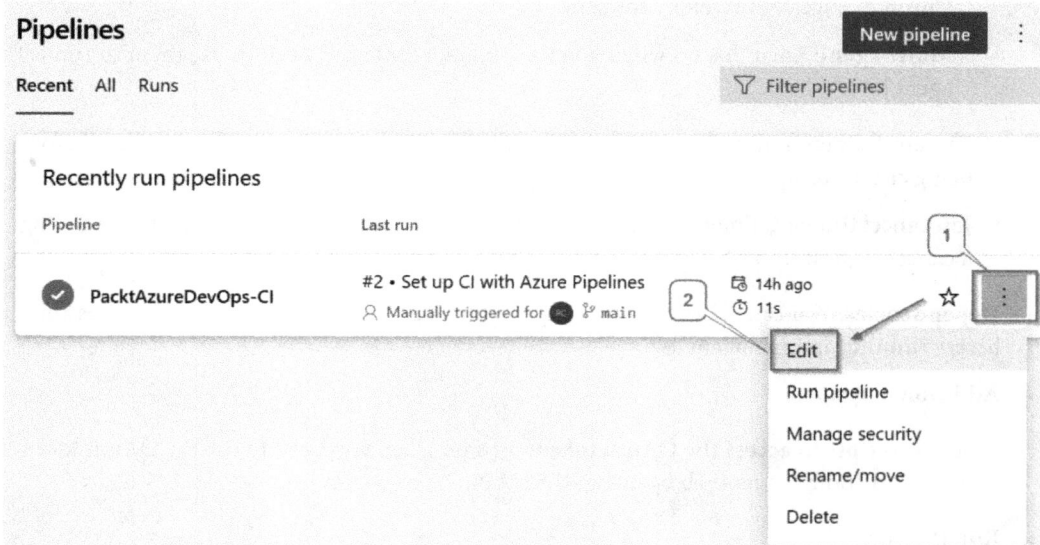

Figure 2.27 – Editing a build pipeline

2. Click on **Triggers** and update the following information:

    *   **Enable continuous integration**: Turn this on

    *   **Branch filters | Type: Include | Branch specification: main**

    If you push your code on the main branch, a build pipeline will run automatically; this is called enabling continuous integration:

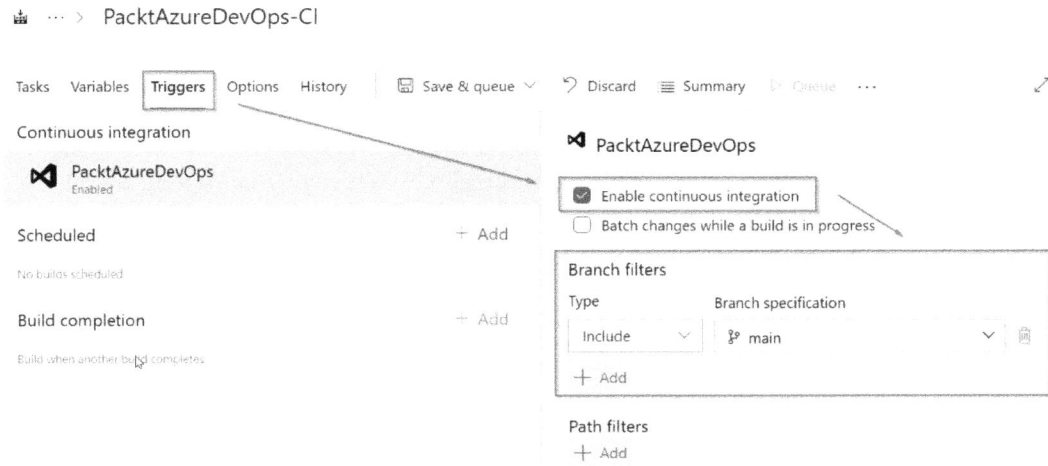

Figure 2.28 – Turning on continuous integration

In this section, you learned how to enable automatically running a build pipeline after we push the code. This will reduce the daily workload to manually build your code whenever you push it to the source code repository.

The following section will teach you how to create a stage to make a task group for deployment environments.

Creating stages

This section will teach you how to create a **stage** using the classic editor – that is, create a task group for environmental purposes such as development, non-functional testing, and production. Each group has jobs and tasks to run for each environment. Using many stages is advantageous when we have many environments to deploy our applications. This is because if we use only one stage by default for the development and production environment, if the development pipeline fails, it will continue to run the production pipeline and also fail. Let's look at how we can create a stage:

1.   Click on **Releases** | **New pipeline**:

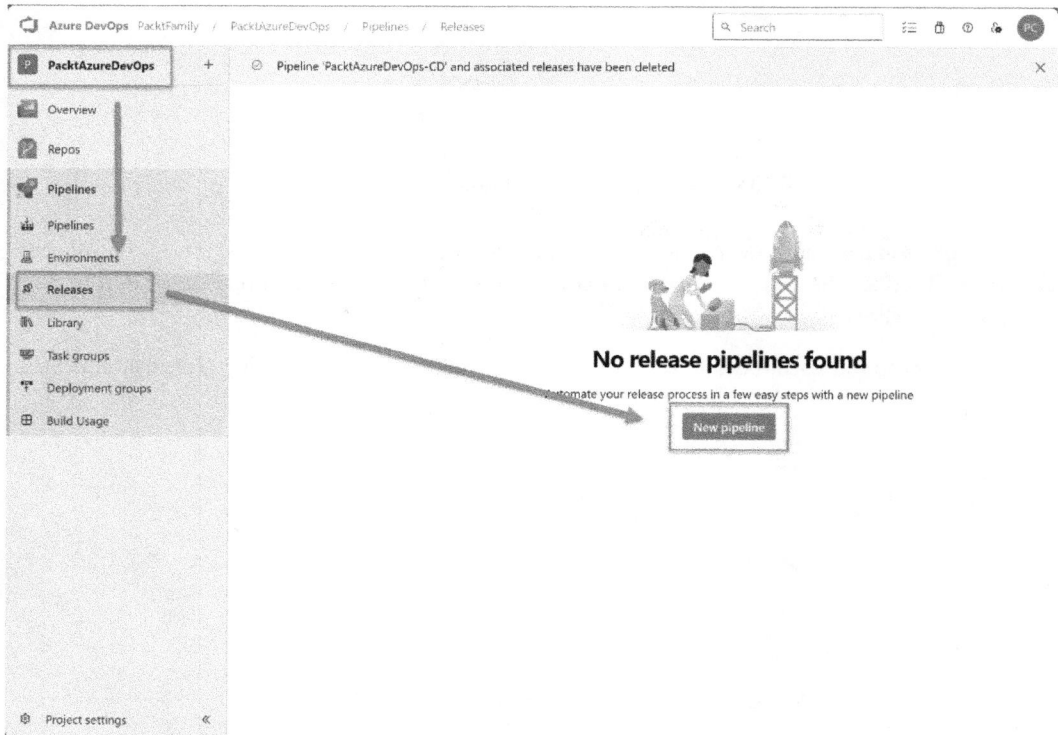

Figure 2.29 – Creating a new pipeline

2. Enter a stage name and click on **Save**:

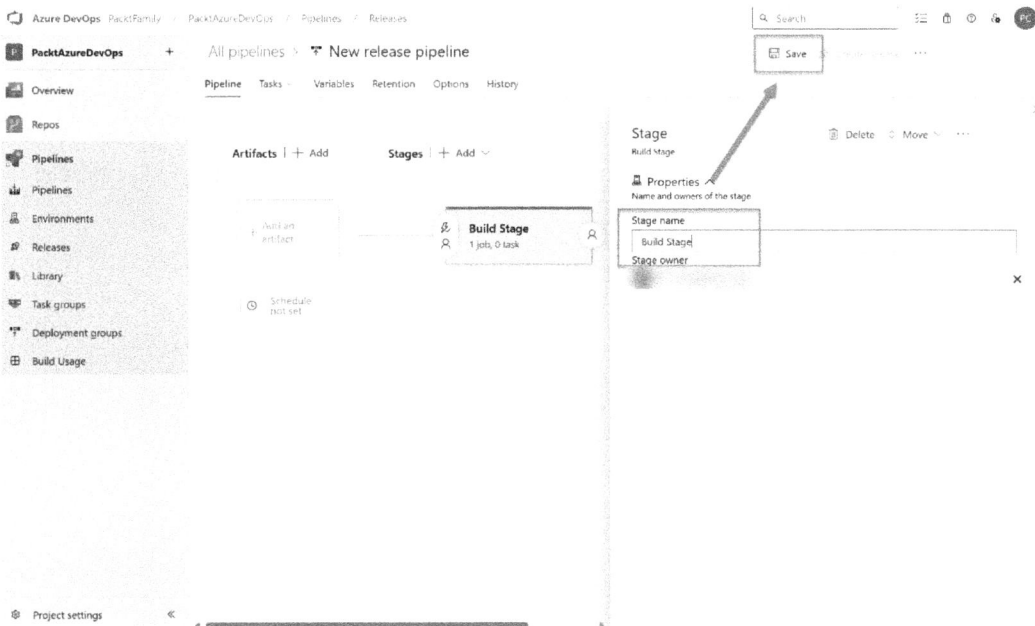

Figure 2.30 – Edit stage properties

3. Select the relevant repository folder. In our case, it is an Azure Repos folder, the root folder for your code and the Azure pipeline file:

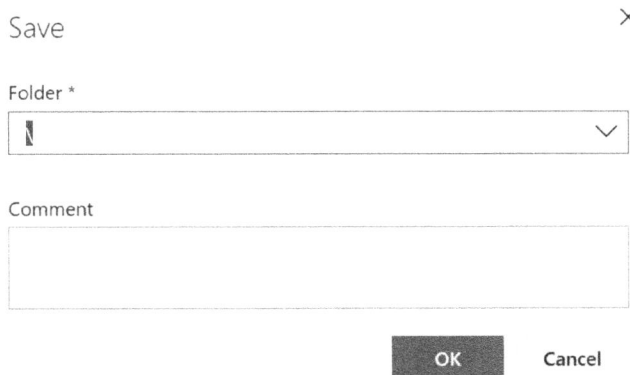

Figure 2.31 – Confirming a pipeline

Now that you've learned how to create a stage with an empty job, let's wrap up this chapter.

Summary

In this chapter, you learned the core features needed to create a pipeline. You learned how to create a build pipeline that includes jobs and tasks. You also learned how to make a trigger to filter any branch to run a build agent. These triggers, combined with jobs and tasks, form the cornerstone of any effective CI/CD pipeline, enabling automated, efficient, and reliable software delivery.

In the next chapter, you will learn how to enhance a build pipeline using YAML and run it on an agent.

3
Setting Variables, Environments, Approvals, and Checks

In the previous chapters, we created our first build pipeline by creating jobs and tasks for it and made a trigger to run a build pipeline automatically. This chapter will cover the next step in creating release pipelines. By the end of this chapter, you will have learned how to create a release pipeline so that you can deploy an application on Azure. This involves setting up a variable group library and learning to add and use secret files for specific use cases, such as mobile application deployment.

We will cover the following topics in this chapter:

- Creating a service connection for Azure resources
- Creating a variable group library
- Uploading and managing a secret file
- Creating a release pipeline.

Creating a service connection for Azure resources

This section will teach you how to create a service connection so that you can release your application on Azure resources. Before creating a service connection, you need to provide an Azure credential that can be obtained from the Azure portal.

Exploring Azure app registration

App registrations is the portal section where you can obtain Azure credentials, allowing Azure Pipelines to deploy applications to Azure resources. You can create an Azure app registration by performing the following steps:

1. Navigate to `https://portal.azure.com` | **Microsoft Entra ID**.
2. Click on **App registrations** and then + **New registration**:

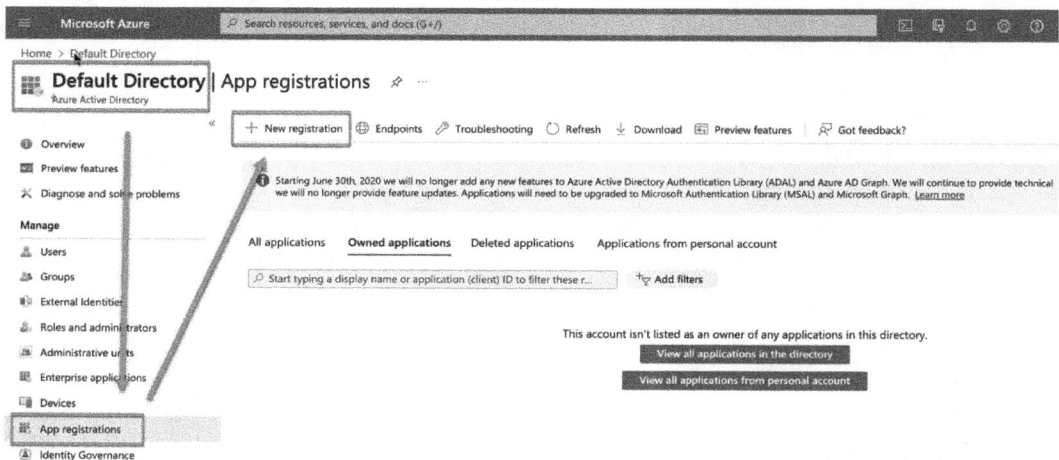

Figure 3.1 – The App registrations page

3. Click on the **Register an application** option. At this point, you need to provide the name of the registered application, after which you can choose from four options regarding account type. The first is for a single tenant, which means only a single identity in one Azure account. The second option is for multiple identities in one Azure account. The third option is for if there are multiple identities, including personal Microsoft accounts in one Azure account. The last option is only for personal Microsoft accounts in one Azure account. Once you've chosen the relevant option, click **Register**:

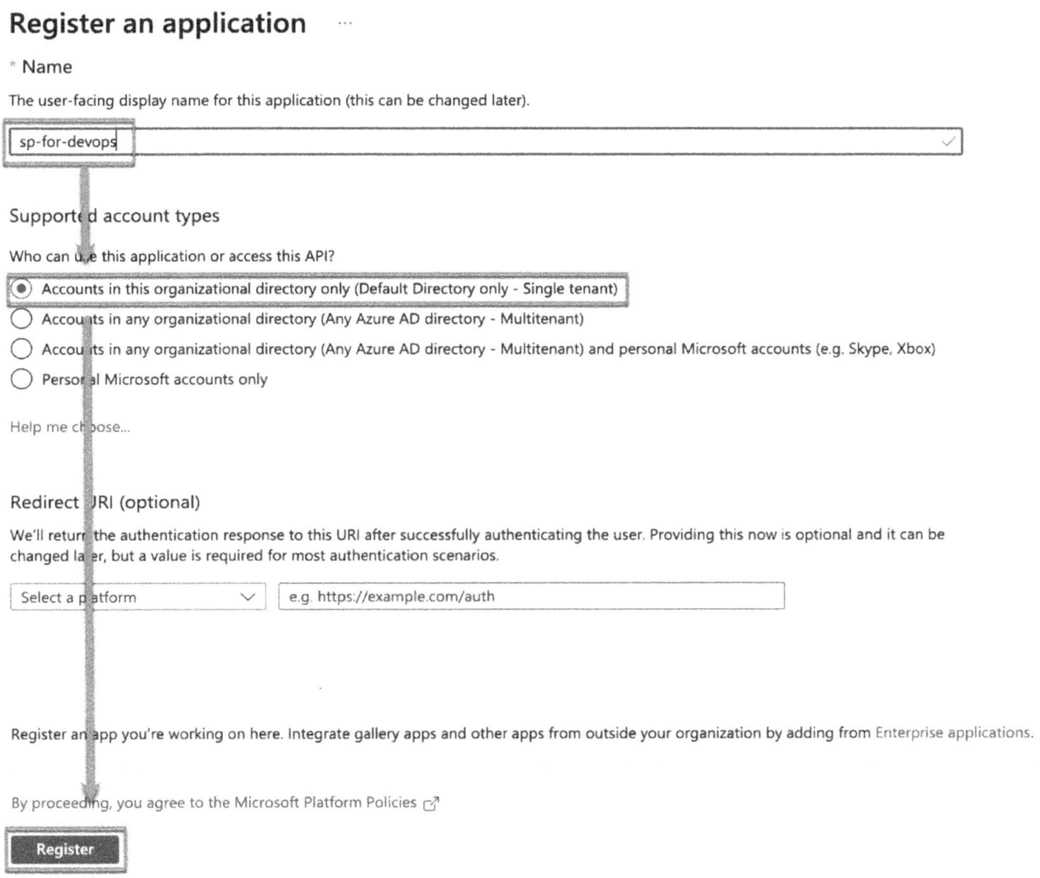

Register an application   ...

* Name

The user-facing display name for this application (this can be changed later).

sp-for-devops

Supported account types

Who can use this application or access this API?

(•) Accounts in this organizational directory only (Default Directory only - Single tenant)

◯ Accounts in any organizational directory (Any Azure AD directory - Multitenant)

◯ Accounts in any organizational directory (Any Azure AD directory - Multitenant) and personal Microsoft accounts (e.g. Skype, Xbox)

◯ Personal Microsoft accounts only

Help me choose...

Redirect URI (optional)

We'll return the authentication response to this URI after successfully authenticating the user. Providing this now is optional and it can be changed later, but a value is required for most authentication scenarios.

Select a platform	∨	e.g. https://example.com/auth

Register an app you're working on here. Integrate gallery apps and other apps from outside your organization by adding from Enterprise applications.

By proceeding, you agree to the Microsoft Platform Policies 🗗

Register

Figure 3.2 – The Register an application page

4. After registering an app ID, navigate to **Certificates & Secrets** to create a secret. Go to **+ New client secret** | **Add a client secret**. Provide the description and expiry date and click **Add**:

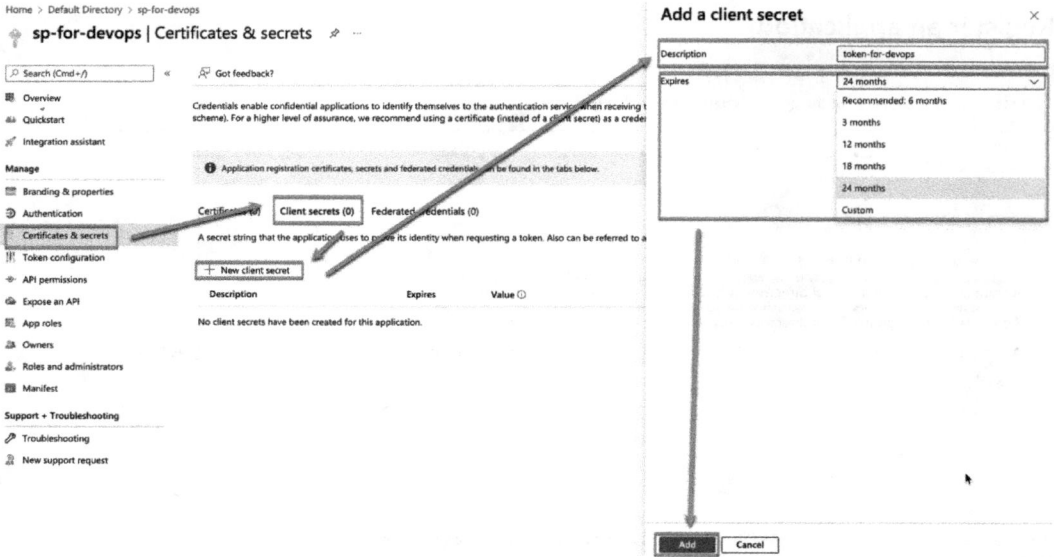

Figure 3.3 – Certificates and secrets

5. Do not forget to copy the *secret* value – it will disappear, and you cannot recover it after you close this page:

Figure 3.4 – Client secrets

6. Copy the highlighted information to prepare for creating a service connection:

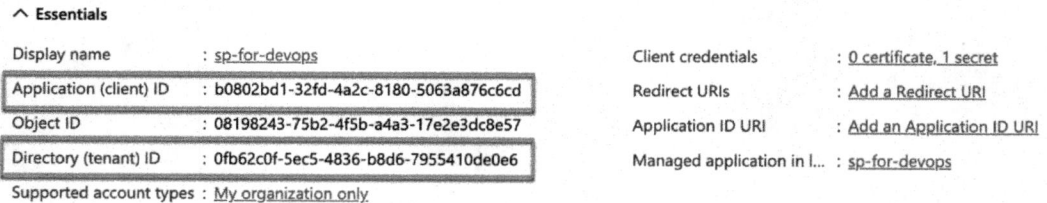

Figure 3.5 – App registration overview

In the following section, we will use this information to create a service connection.

Creating a service connection

To deploy an application on Azure, you need to create a service connection, which is a service account that allows you to access the Azure resource. To do this, follow these instructions:

1. After creating an app registration from the Azure portal, navigate to the Azure DevOps page at `https://dev.azure.com/` and click **Sign in**:

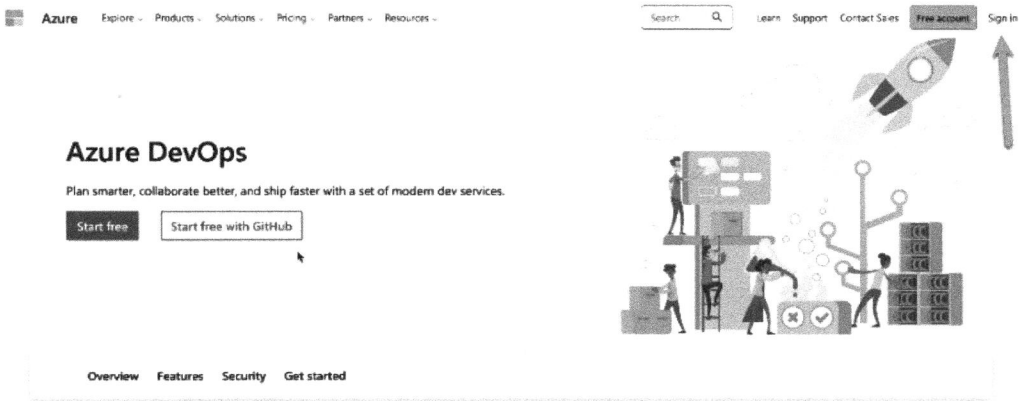

Figure 3.6 – Sign-in page

2. Click on **Project settings | Service connections > Create service connection**:

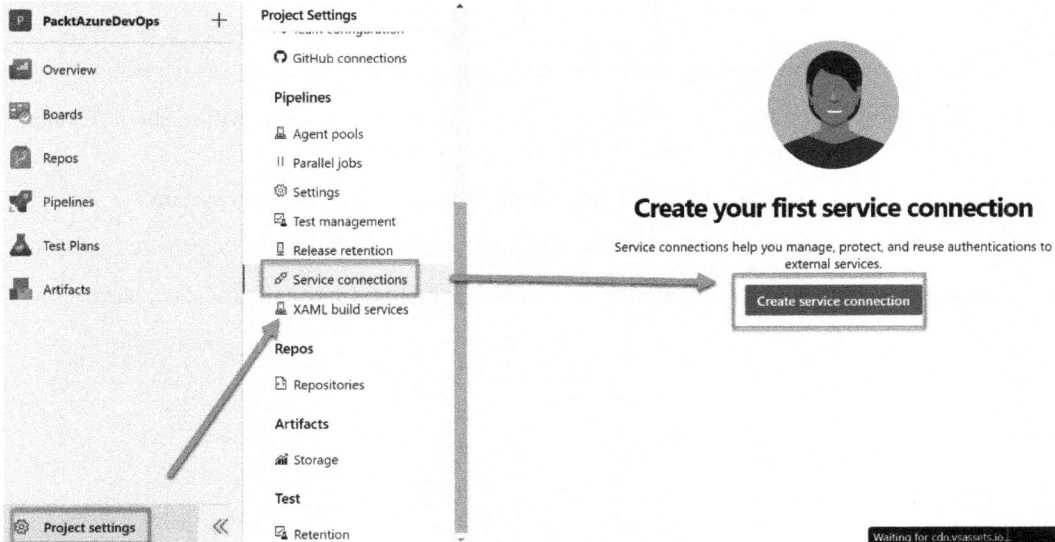

Figure 3.7 – Creating a service connection

3.   Select **Azure Resource Manager** and click **Next**:

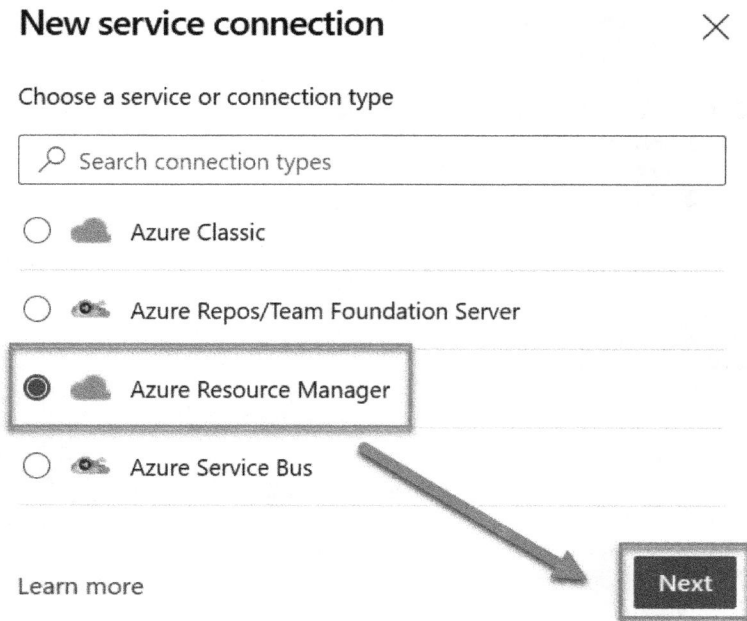

Figure 3.8 – Selecting a service connection type

4.   For **Authentication method**, there are four options to choose from:

- The first option involves automatically finding service principles in all Azure resources.

- The second option is a manual method where you can enter all information about the app ID. This option allows you to easily connect to Azure resources.

- The third option is used for an existing identity that is used for another system.

- The final option involves exporting the public profile from the Azure portal page and using it.

For this example, select the second option, **Service principal (manual)**, and click **Next**:

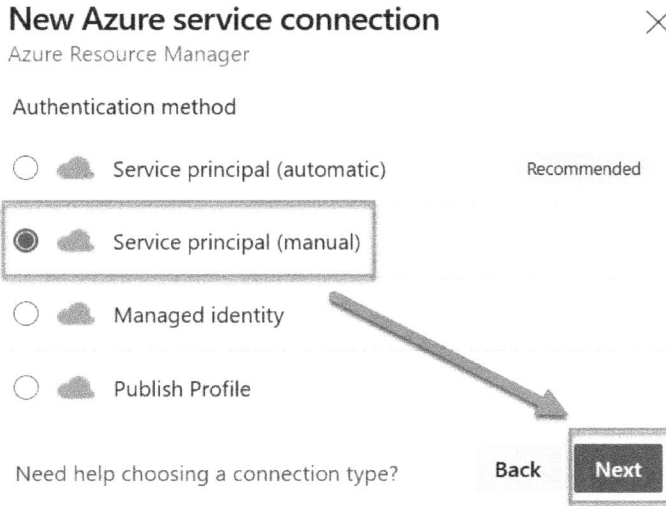

Figure 3.9 – Choosing an authentication method

5. Fill in the required fields, as follows, and click **Verify and save**:

*   **Environment**: Azure Cloud
*   **Scope Level**: **Subscription** (there are three levels; we're using this option because we need to limit the scope of permission to only the subscription level for risk management)
*   **Subscription Id**: <Check on subscription menu>
*   **Subscription Name**: <Check on subscription menu>
*   **Service Principal Id**: <Client id of App registration>
*   **Credential**: **Service principal key**
*   **Service principal key**: <Secret of App registration>
*   **Tenant ID**: <Tenant ID of App registration>
*   **Service connection name**: sp-for-devops
*   **Security**: Turn **Grant access permission to all pipelines** on.

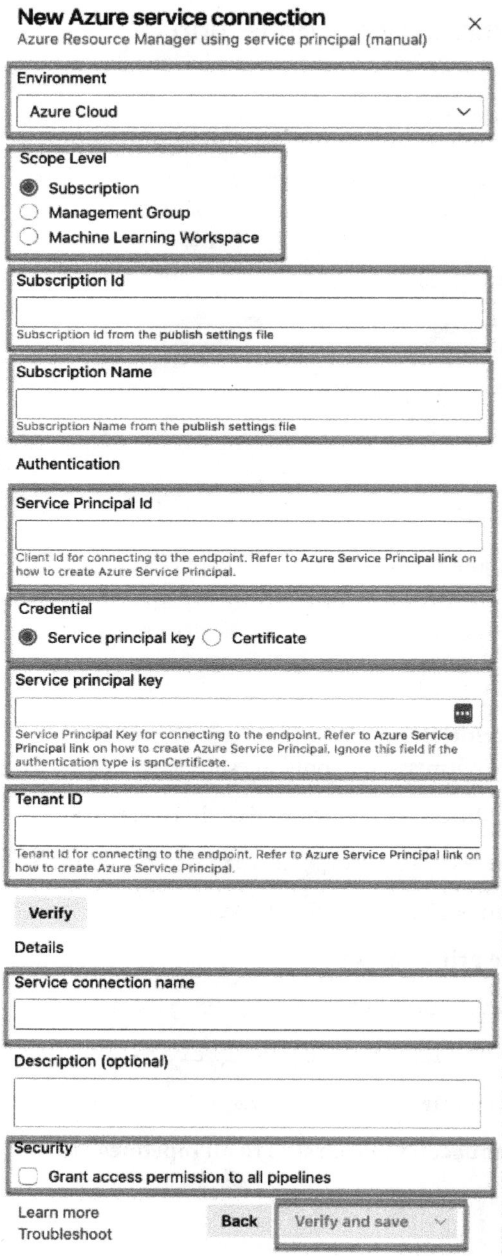

Figure 3.10 – Enter Azure service connection details

6.   You can now view the new service connection:

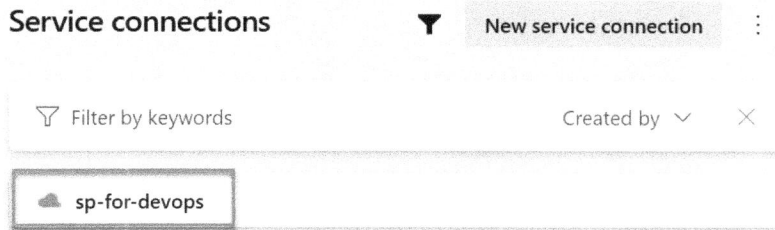

Figure 3.11 – List of all service connections

In this section, you learned how to create a service connection for an Azure resource connection. You will use this when you create a release pipeline for deploying an Azure application. In the next section, you will learn how to create global variables and secret files for use on all Azure pipelines in a project.

Managing global variables and secret files

Most projects will use the same value for creating release pipelines, such as the name of the Azure service connection. This section will teach you how to create the variable group and secret files. These resources are essential for sharing common values across multiple pipelines, and include, for instance, a username and password for deploying applications to Microsoft Azure.

Creating a variable group library

You need to create a global variable to link to all pipelines if that pipeline needs to use an Azure service connection. Using variable groups reduces the chances of making mistakes and duplicating values across many pipelines. When you need to update the values, you can do so in a single centralized location instead of throughout all pipelines.

You can follow these steps when you need to create variables you wish to share for all pipelines:

1. Navigate to your project and click on **Pipelines** | **Library** | **Variable group**:

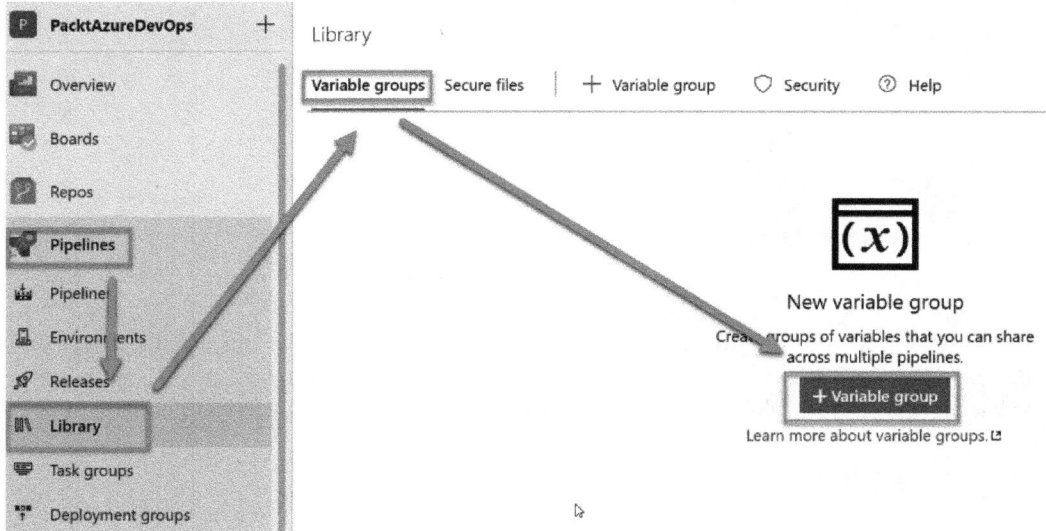

Figure 3.12 – Creating a variable group

2.   Fill in all the required fields; specify **Variable group name** and **Name** with the values that you would like to share across all pipelines and click **Save**:

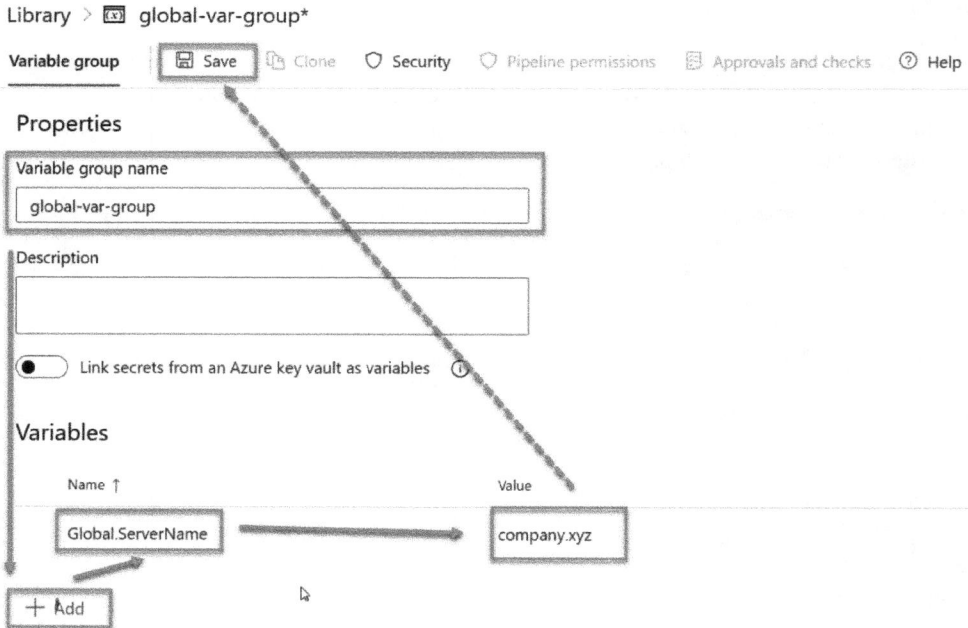

Figure 3.13 – Entering variable group details

3. Edit the existing pipeline that you need to link to a variable group. Click the **Edit** option for the existing pipeline and navigate to **Variables** | **Variable groups** | **Link variable group**:

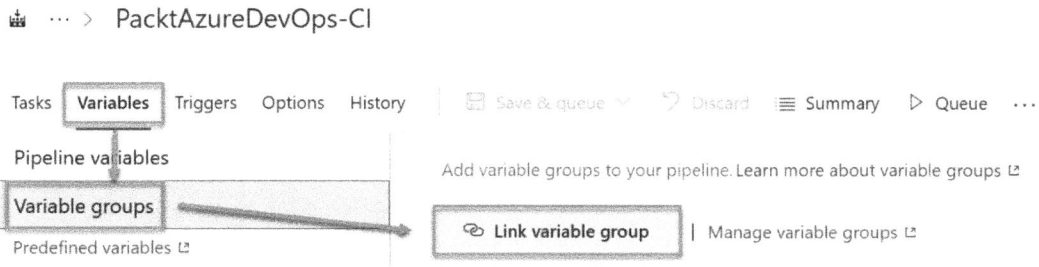

Figure 3.14 – Selecting variable groups

4. Select the variable group you just created and click **Link**:

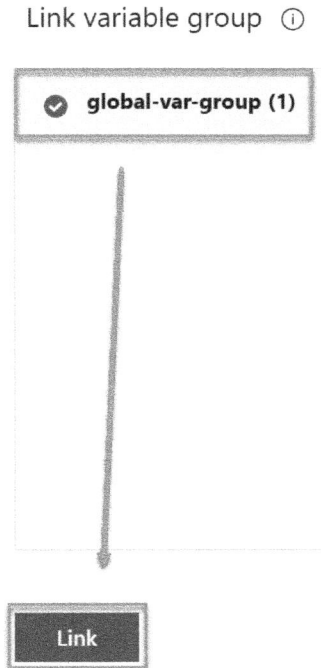

Figure 3.15 – Link variable group

5.   Review all the values that have been assigned to your variable group and click **Save & queue**:

Figure 3.16 – Reviewing and saving the variable group to the pipeline

In this section, you learned how to create a variable group to share for all pipelines. You also learned how to link a variable group to the existing pipeline. In the next section, you will learn how to upload secret files and connect them to your pipelines.

Uploading and selecting a secret file

Secret files typically contain sensitive information, such as signing certificates, SSH keys, license files, or mobile provisioning profiles. In some cases, you are responsible for generating such files and in other cases, these files are generated on other platforms. You then must download the file from that platform and make it available to your CI/CD pipelines.

For example, to deploy a mobile application such as an iOS application, you must first generate a **provisioning profile** that contains information about who is developing the iOS application, download it, and then link it to the pipeline.

To do this, follow these steps:

1.   Go to **Pipelines | Library | + Secure file**:

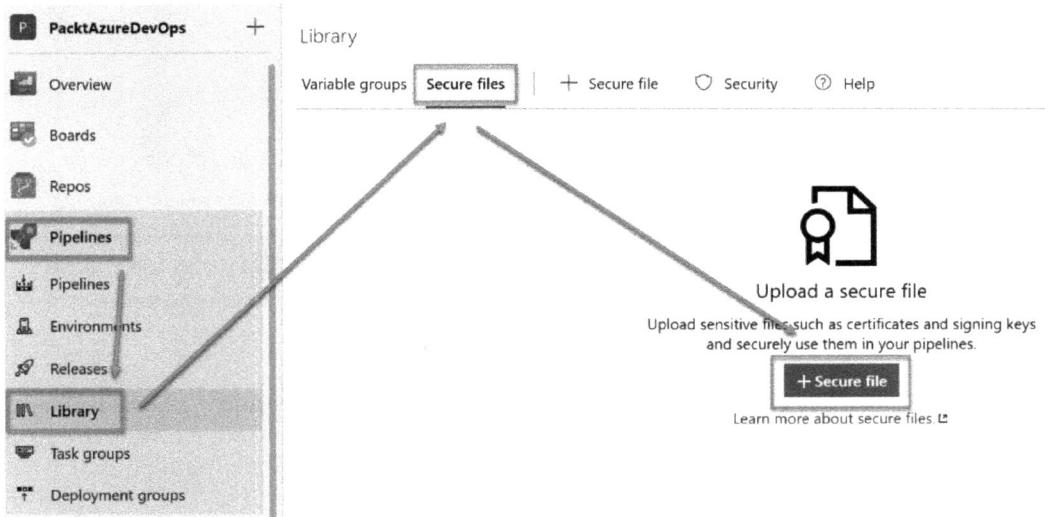

Figure 3.17 – Secure file

2. Click **Browse…** to select the `*.mobileprovision` file that you need to keep as a secret file and click **OK**:

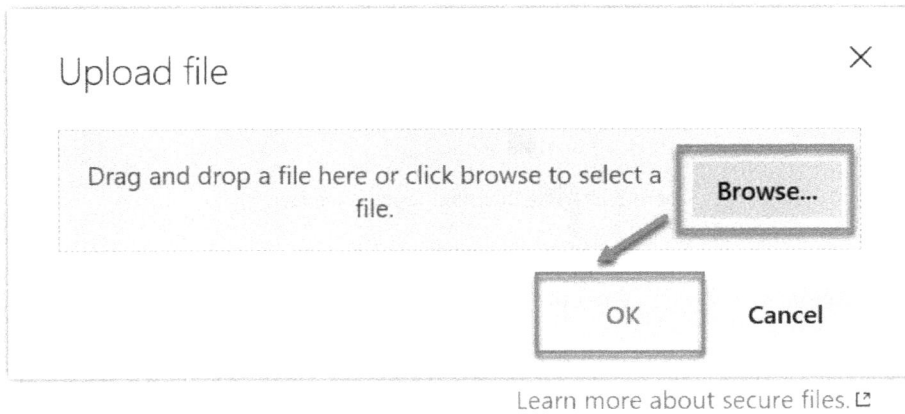

Figure 3.18 – Upload file

3. View the secure file that you uploaded:

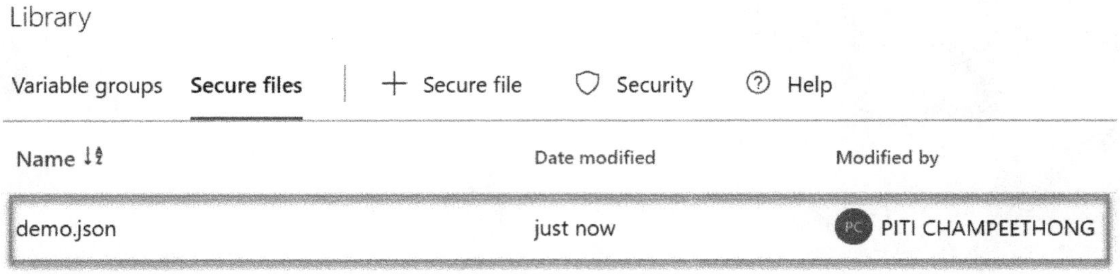

Figure 3.19 – Viewing secure files

4. Click on + on the relevant job in the pipeline, type `Secure`, and click **Add** on the **Download secure file** task:

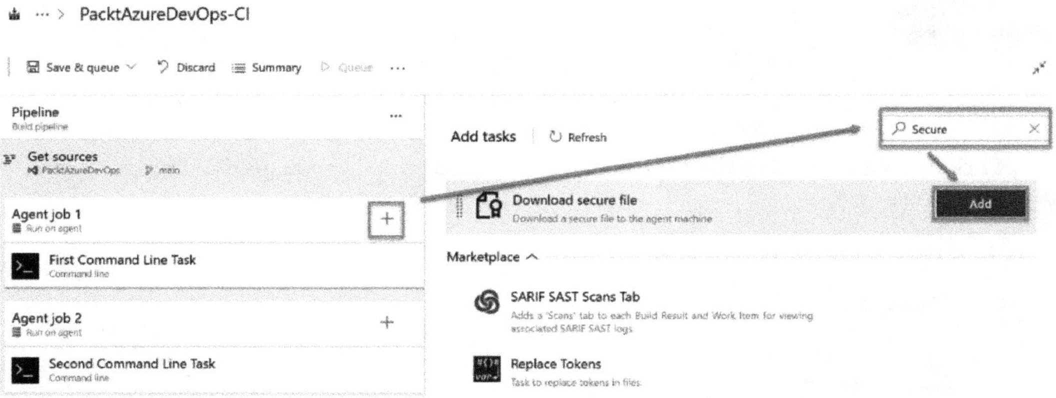

Figure 3.20 – The Download secure file task

5. In the previous steps, we added the secret file in the global section. We now need to download it when we use it. To do this, enter the relevant details:

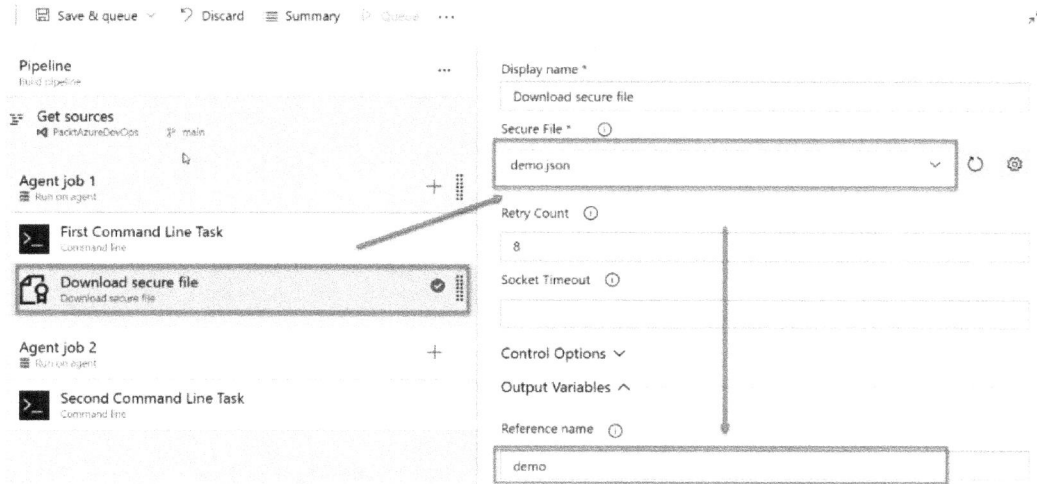

Figure 3.21 – Entering the relevant details

6. Click + on the relevant job in the pipeline, type Command line, and click **Add** on the **Command line** task:

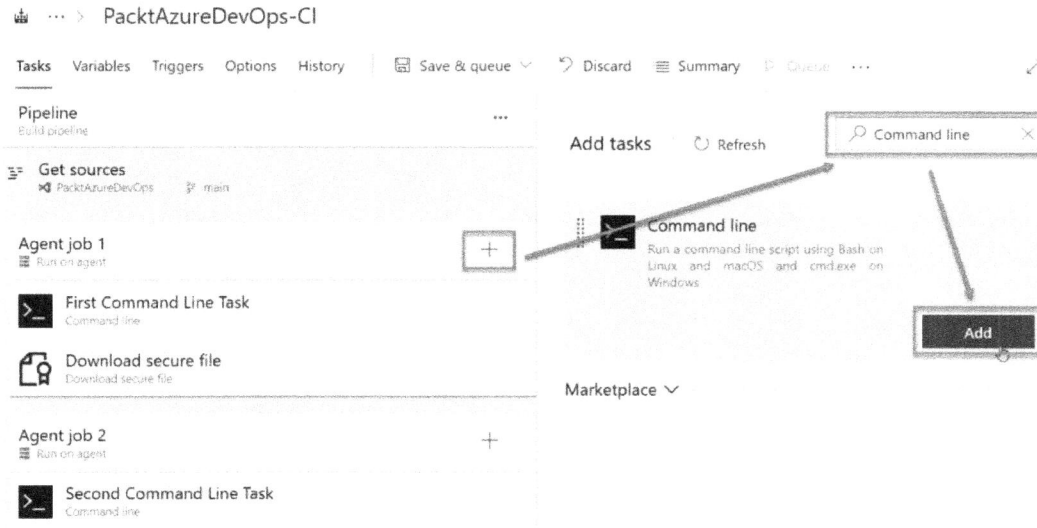

Figure 3.22 – Adding the Command line task

7. Enter the relevant **Command line** details: set **Display name** to `Display secure file path` and for **Script**, enter `echo $(demo.secureFilePath)`. Click **Save & queue**:

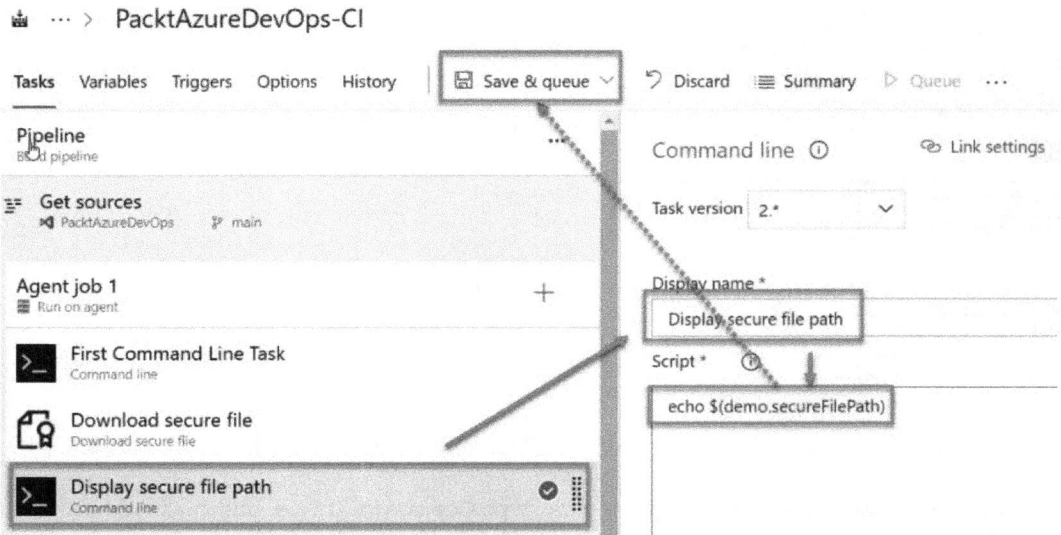

Figure 3.23 – Entering details for the Command line task

8. View the job to see the **Display secure file** log:

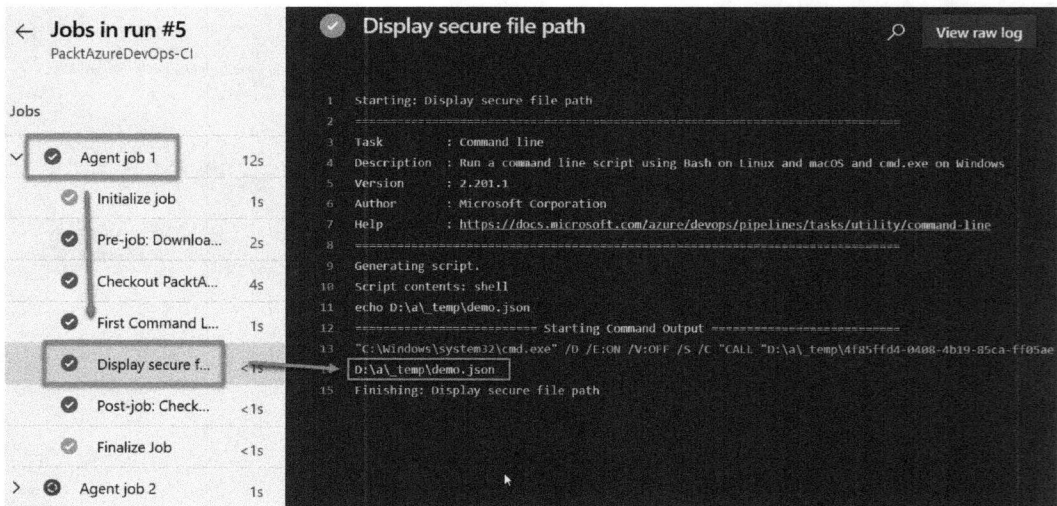

Figure 3.24 – Viewing the secure file path

> **Why use secure files?**
>
> These files will be stored securely and encrypted in Azure DevOps for you to use in your pipelines, minimizing the chances for them to be misplaced or misused by your team.

This section taught you how to create secure files and connect them to pipelines. You also learned how to download secure files to pipelines and display them on the command line. In the next section, you will learn how to create a release pipeline that includes both a variable group and a secure file.

Creating a release pipeline

This section will teach you how to create a release pipeline to deploy the artifact that's received from a build pipeline.

A **release pipeline** also includes a secure file and a variable from a variable group of a library.

To create such a pipeline, follow these steps:

1. Navigate to **Pipelines | Releases | New pipeline**:

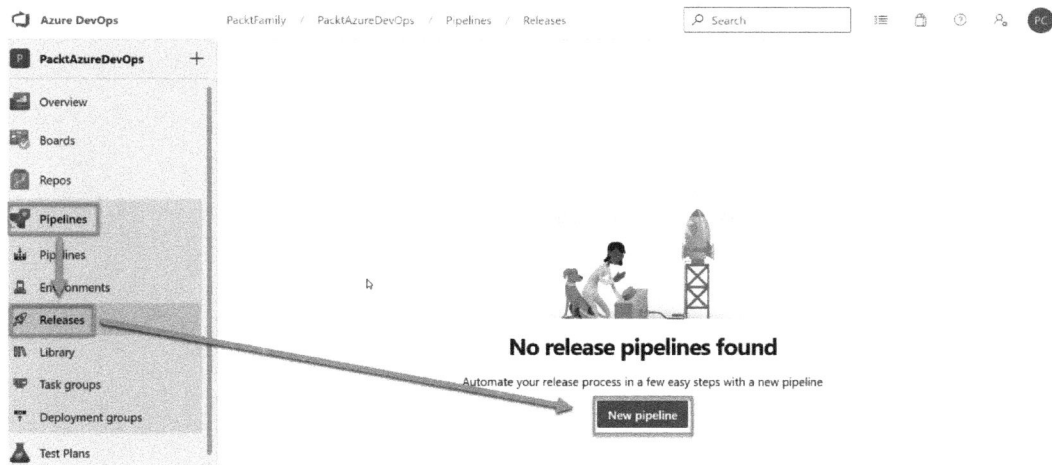

Figure 3.25 – New pipeline

2. Click on the relevant pipeline, which in this case is **release-app-dev**. Navigate to **Pipeline | Artifacts | + Add**:

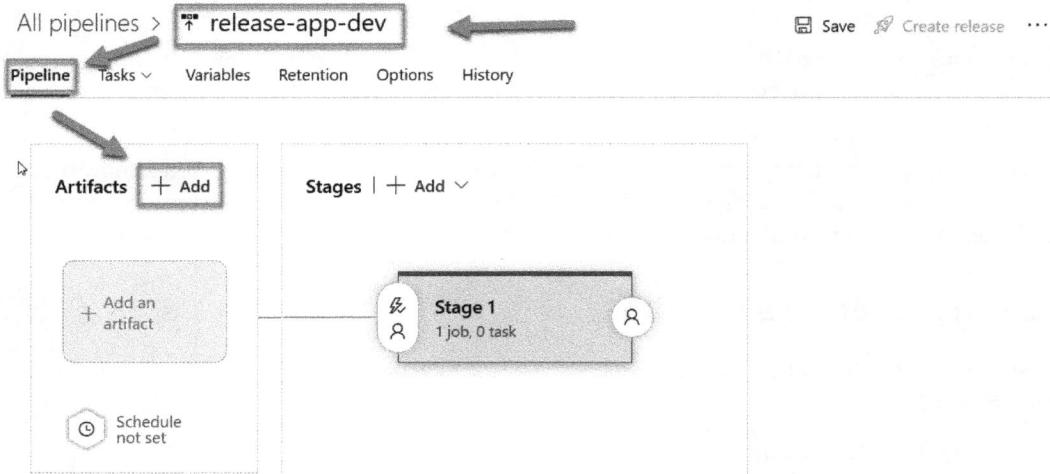

Figure 3.26 – Adding artifacts

3.   Click **Build** and select **PacktAzureDevOps-CI**, then click **Add**:

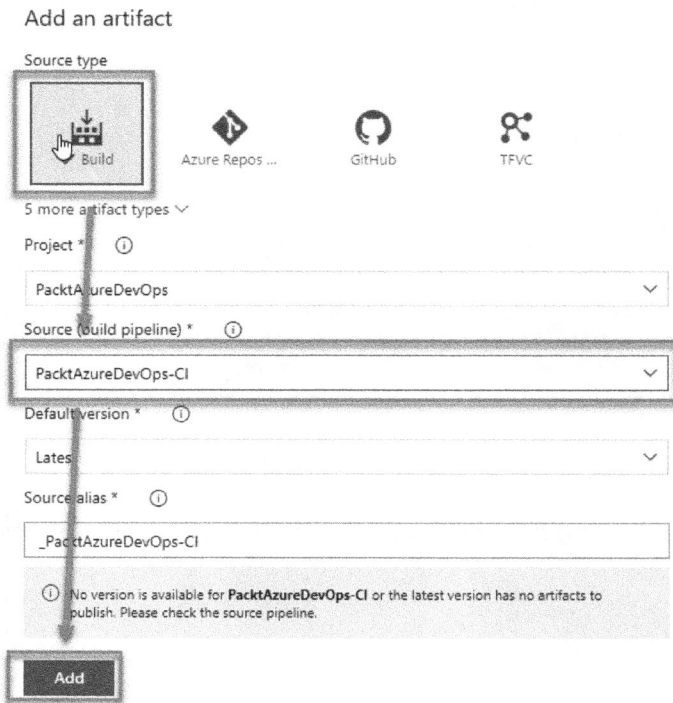

Figure 3.27 – Entering artifact data

4. Go to **Variables** | **Variable groups** | **global-var-group** | **Link**:

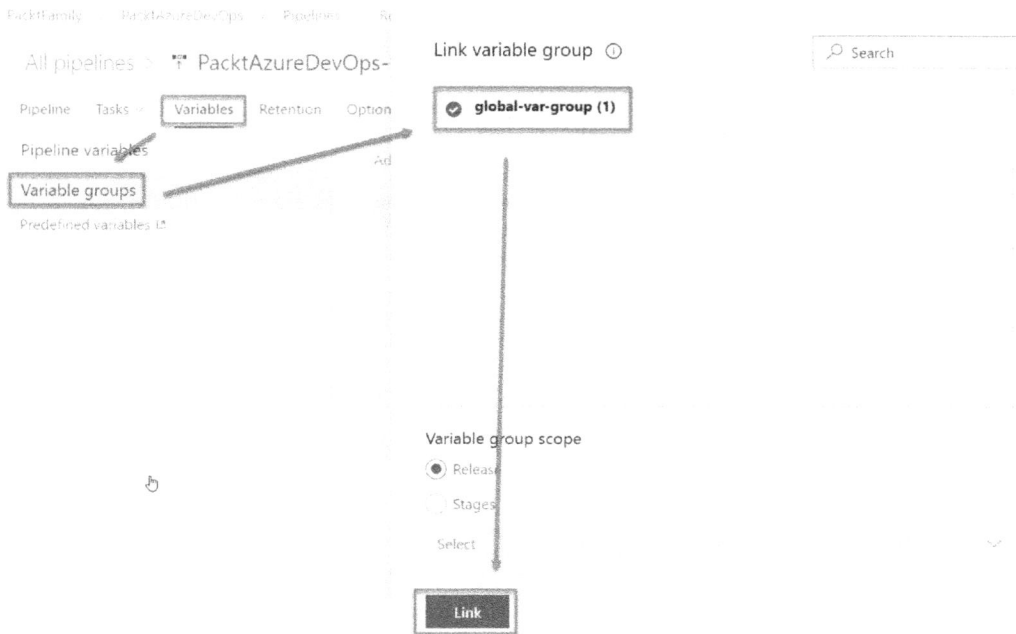

Figure 3.28 – Link variable group

5. Expand **global-var-group (1)** to see all the associated variables:

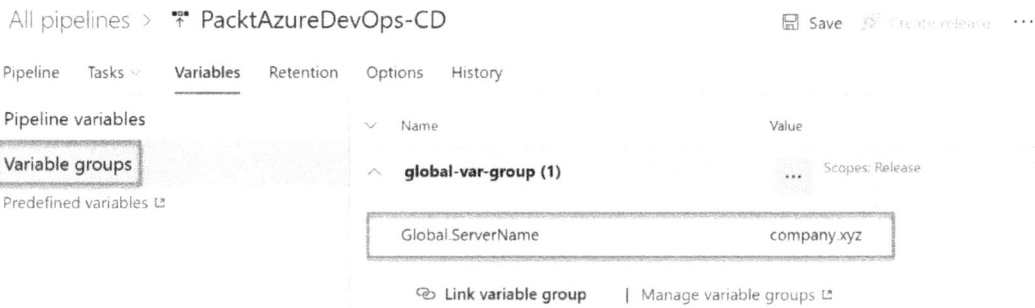

Figure 3.29 – Reviewing the variables

6. Click **Download secure file**, then select **demo.json** and enter demo under **Reference name**:

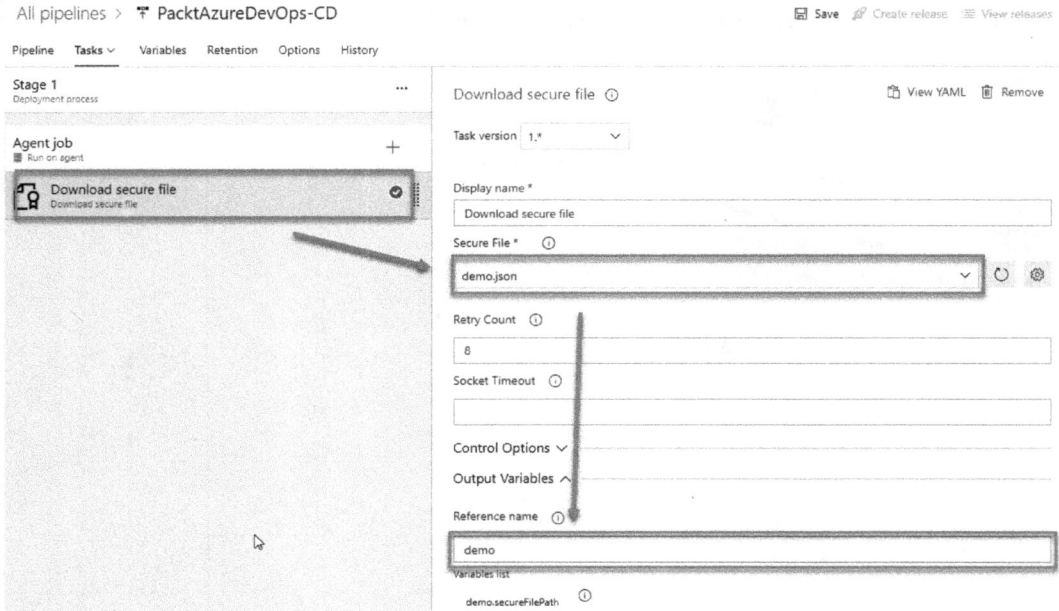

Figure 3.30 – Adding a Download secure file task

7.   Click on the **Command line** task and enter the relevant information, as follows:

- **Display name**: Display Secure file & Variables

- **Script**:

```
echo $(demo.secureFilePath)
echo "==============="
echo $(Global.ServerName)
```

This can be seen in the following screenshot:

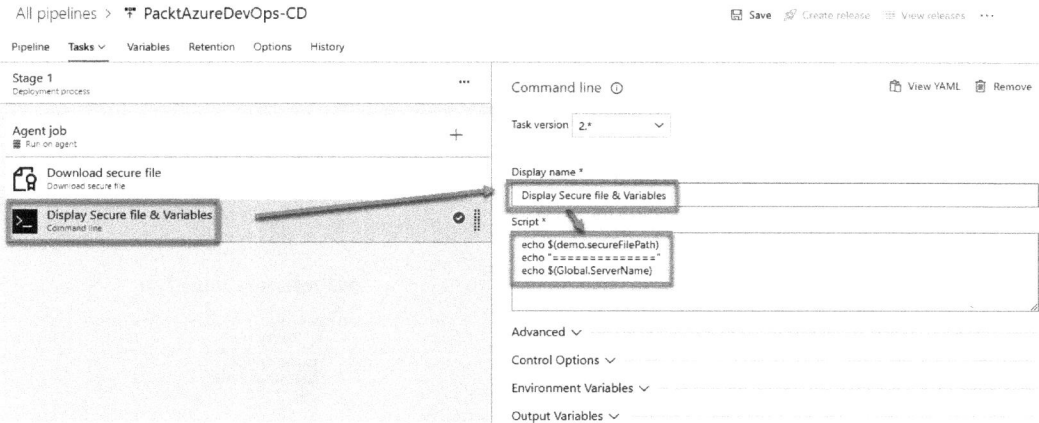

Figure 3.31 – Adding a Command line task

8. Click on **Agent job**, then select the following options:

    ▪ **Agent pool: Azure Pipelines**

    ▪ **Agent Specification: ubuntu-latest**

    Then click **Save**:

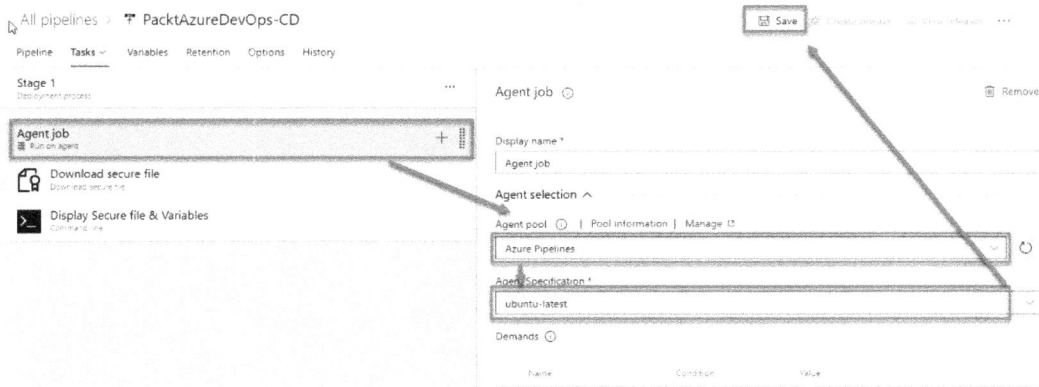

Figure 3.32 – Agent job properties.

9. To create the release pipeline, navigate to **Releases** | **PacktAzureDevOps-CD** | **Releases** | **Create a release**:

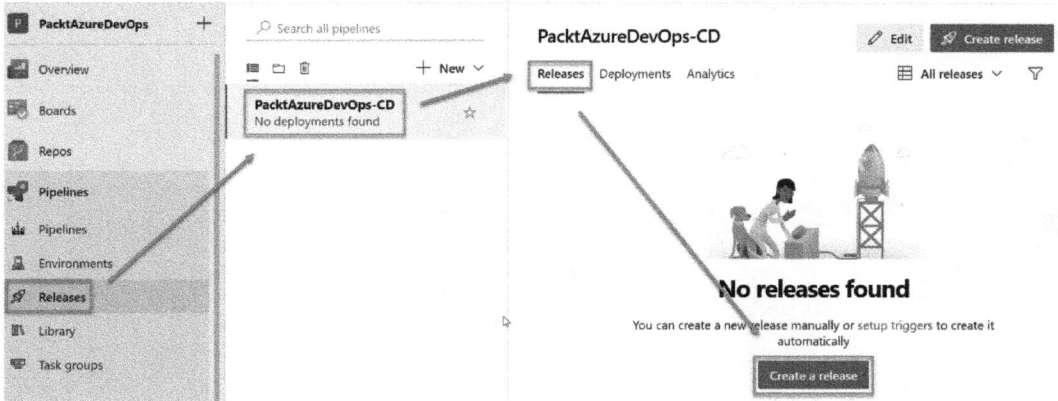

Figure 3.33 – Create a release

10. You can retain the default values. You can also enter information to describe the purpose of the release pipeline in the **Release description** box and click **Create**:

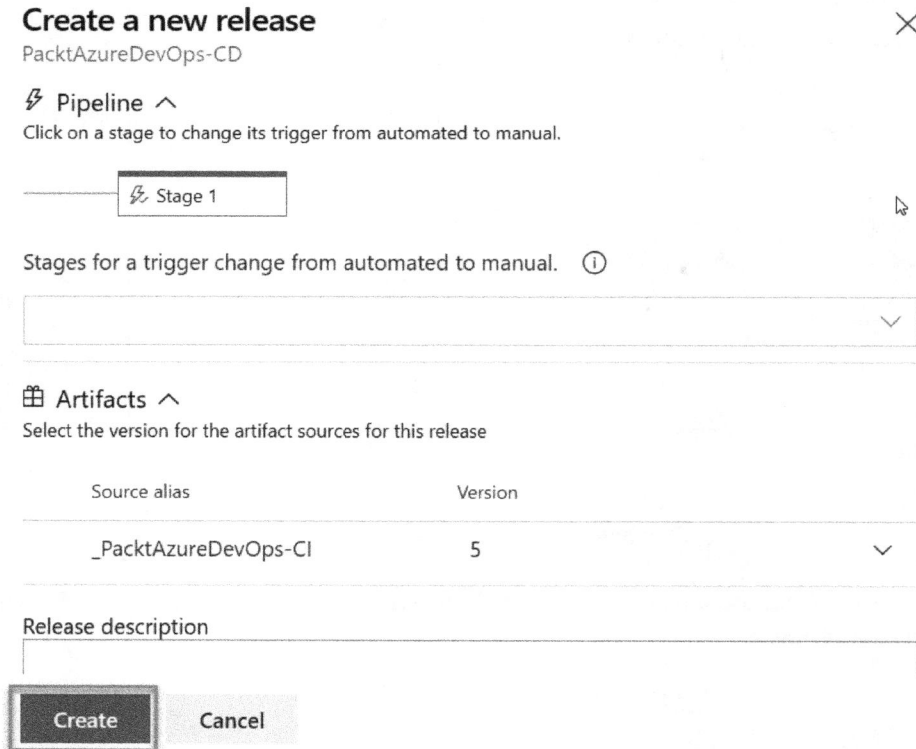

Figure 3.34 – Confirming the creation of a release pipeline

The default values in the preceding screenshot have different meanings; let's break them down:

- Stages with an automated trigger will start as soon as the release is created. With this option, you can stop the stage from starting when the release is created and then start it manually from the portal.

- For **Artifacts**, the latest version available when creating the release will be the default option that's selected. This option gives you the chance to select any other version available, such as in the scenario where you want to perform a rollback by deploying a previous version of the artifact.

11. You can now see the progress of running the release pipeline:

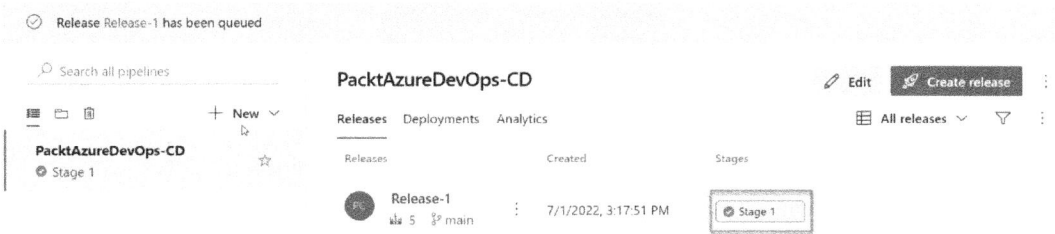

Figure 3.35 – Building a release

12. You can also view the task list by clicking **Display Secure file & Variables**:

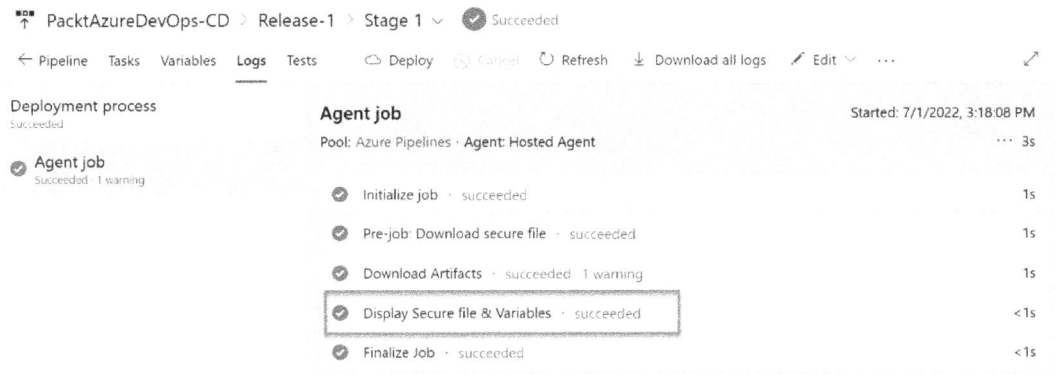

Figure 3.36 – Viewing the release results

The task results will be shown as follows:

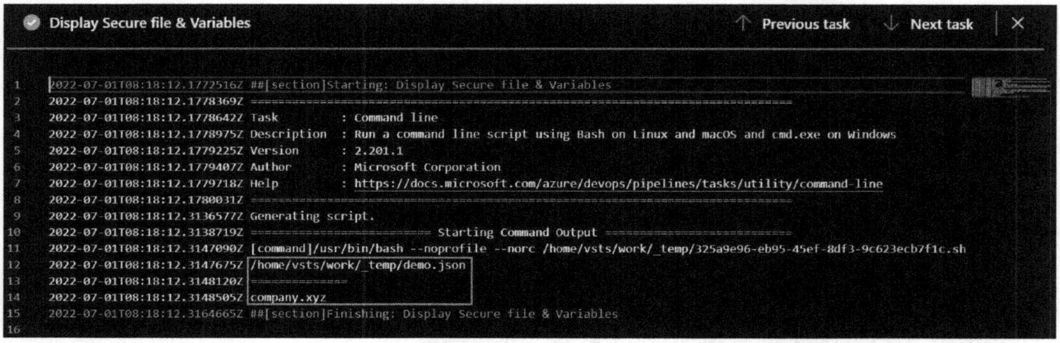

Figure 3.37 – Command line task results

With that, you have learned how to create a release pipeline, including a secure file and a variable group library. You can apply this pattern to any pipeline when creating one.

Summary

This chapter taught you how to create a release pipeline with variable group libraries and upload and manage a secret file. These components not only enhance the organization and security of your pipelines but also provide a foundation for efficient, consistent deployments.

Understanding how to create them will help you orchestrate streamlined, effective interactions between your Azure DevOps pipeline and the Azure ecosystem, a valuable skill for any Azure-powered development project.

In the next chapter, you will learn how to customize a build pipeline using YAML and run it on the agent. This YAML will create an advanced build pipeline rather than creating it on the Azure portal and is where you can create parameters for a build pipeline.

4

Extending Advanced Azure Pipelines Using YAML

We created a build pipeline in the previous chapter by creating jobs, tasks, and triggers. This chapter will teach you how to customize the Azure pipeline using YAML, such as by creating condition statements with variable groups when setting some complex conditions. It also helps to create a flexible pipeline rather than the classic online version. For instance, YAML can do that when you need to deploy mobile applications to both the Google Play Console and the App Store connect simultaneously.

By the end of this chapter, you will have learned about creating a build and release pipeline using YAML. You will also have learned how to clone, export, and import YAML from the classic editor in the Azure DevOps portal.

We will cover the following topics:

- Creating a build pipeline using YAML
- Creating a release pipeline using YAML
- Cloning, exporting, and importing a YAML pipeline
- Complex YAML configurations
- Advantages and limitations of YAML-based pipelines

Let's start by creating a pipeline using the YAML syntax.

Creating a build pipeline using YAML

In this section, you will learn how to build a pipeline using YAML. You will also learn how to view YAML on the Azure DevOps portal and save the YAML file in Azure Repos. To create a build pipeline using YAML, follow these steps:

1. After logging in via the Azure DevOps portal, please select your organization and then navigate to the **Pipelines** page. Click on **New pipeline**:

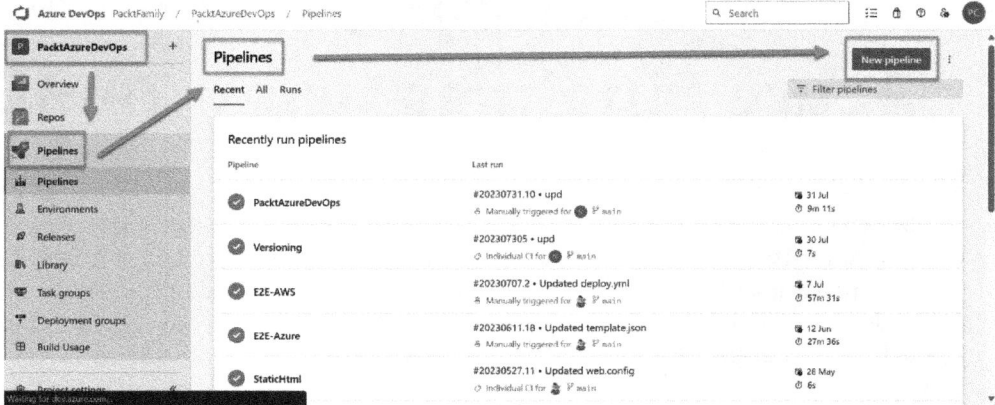

Figure 4.1 – New pipeline

2. Click on **Azure Repos Git**, which is a source code repository for the demo:

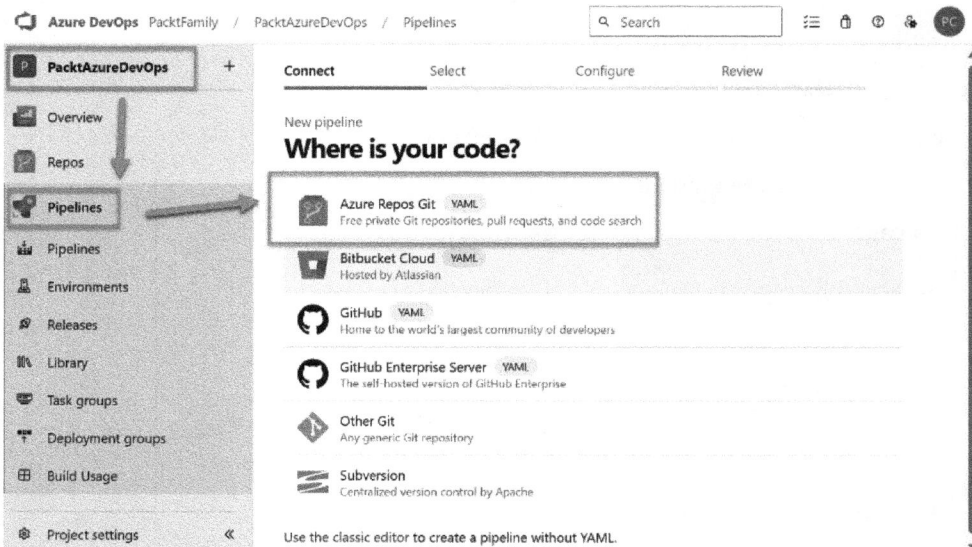

Figure 4.2 – Azure Repos Git

3. Click on the **PacktAzureDevOps** repository:

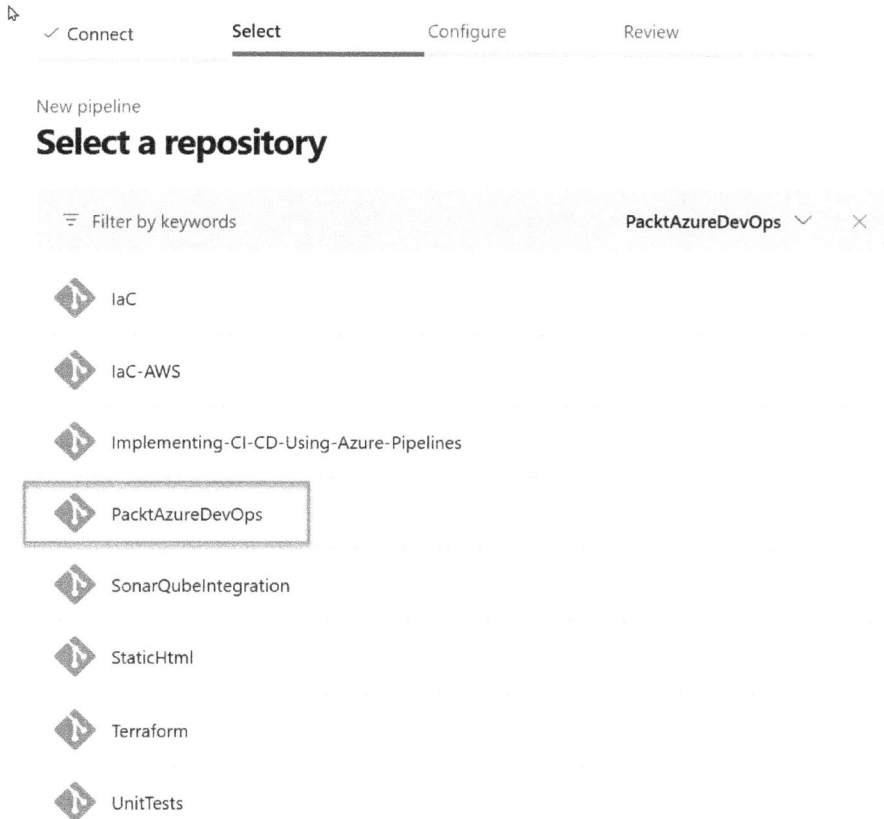

Figure 4.3 – Select a repository

4. If you have existing YAML, you will have to select **Existing Azure Pipelines YAML file**. However, as we are creating a new one here, we will click on **Starter pipeline**:

✓ Connect	✓ Select	Configure	Review

New pipeline

Configure your pipeline

Starter pipeline
Start with a minimal pipeline that you can customize to build and deploy your code.

Existing Azure Pipelines YAML file
Select an Azure Pipelines YAML file in any branch of the repository.

Show more

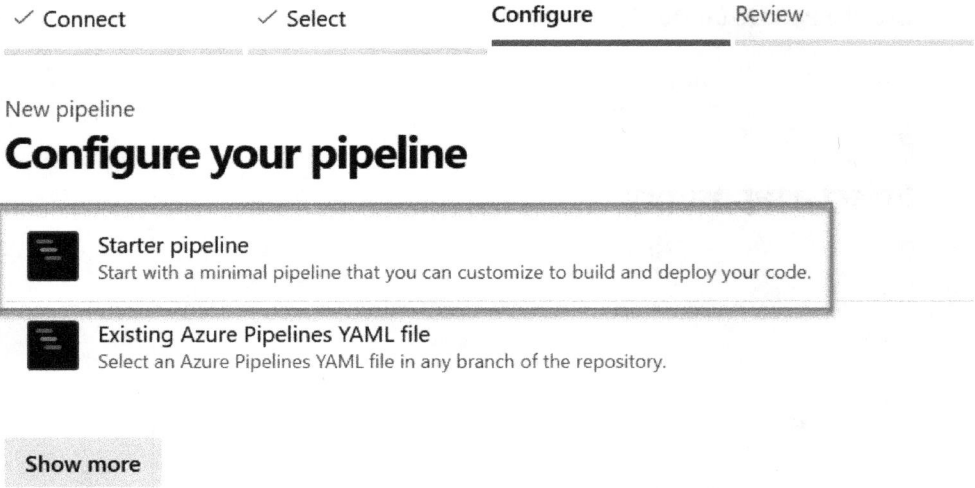

Figure 4.4 – Selecting a starter pipeline

5. Click on **Save**, and you can review a new pipeline in YAML format:

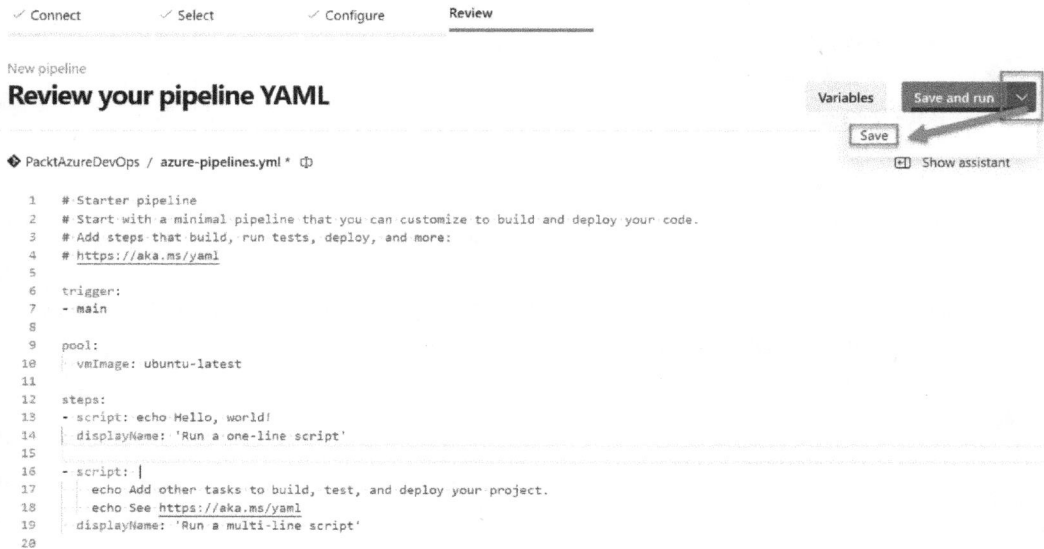

✓ Connect	✓ Select	✓ Configure	Review

New pipeline

Review your pipeline YAML

Variables   Save and run   ▼
Save
⊞ Show assistant

◆ PacktAzureDevOps / azure-pipelines.yml *

```
 1   # Starter pipeline
 2   # Start with a minimal pipeline that you can customize to build and deploy your code.
 3   # Add steps that build, run tests, deploy, and more:
 4   # https://aka.ms/yaml
 5
 6   trigger:
 7   - main
 8
 9   pool:
10     vmImage: ubuntu-latest
11
12   steps:
13   - script: echo Hello, world!
14     displayName: 'Run a one-line script'
15
16   - script: |
17       echo Add other tasks to build, test, and deploy your project.
18       echo See https://aka.ms/yaml
19     displayName: 'Run a multi-line script'
20
```

Figure 4.5 – Save pipeline YAML

6. Enter a commit message that helps you remember what you did in the file and select the **Commit directly to the main branch** option. This option will save your file in the main branch:

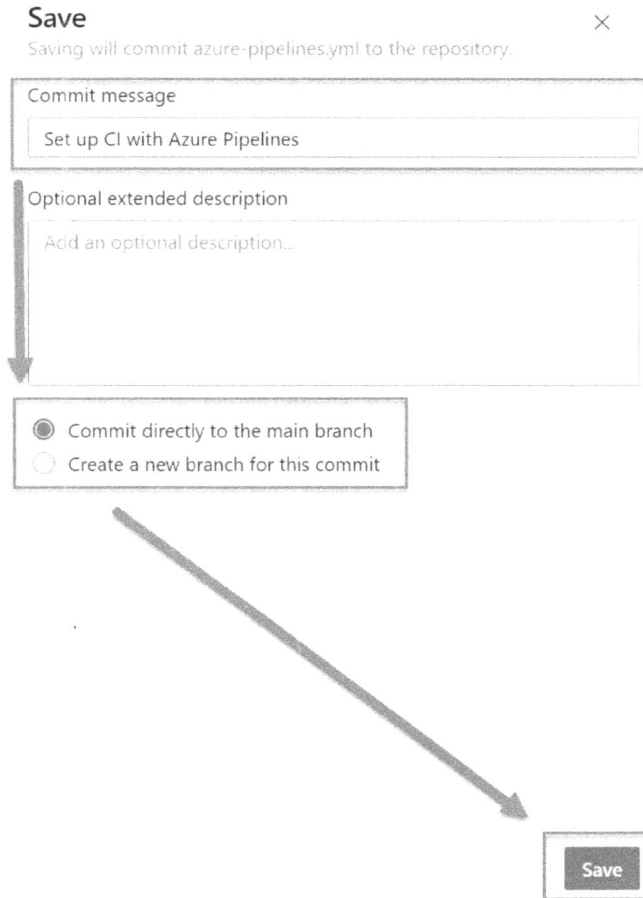

Figure 4.6 – Commit YAML to Azure Repos Git

If you would like to save your file in a new branch, then select **Create a new branch for this commit**. Click on **Save**.

7.   After you click **Save**, you will be taken back to the main dashboard result of the build pipeline:

Figure 4.7 – A build pipeline dashboard

8.   Click on **Run pipeline**:

Figure 4.8 – Run a build pipeline

After that, you can see a summary of the build pipeline results:

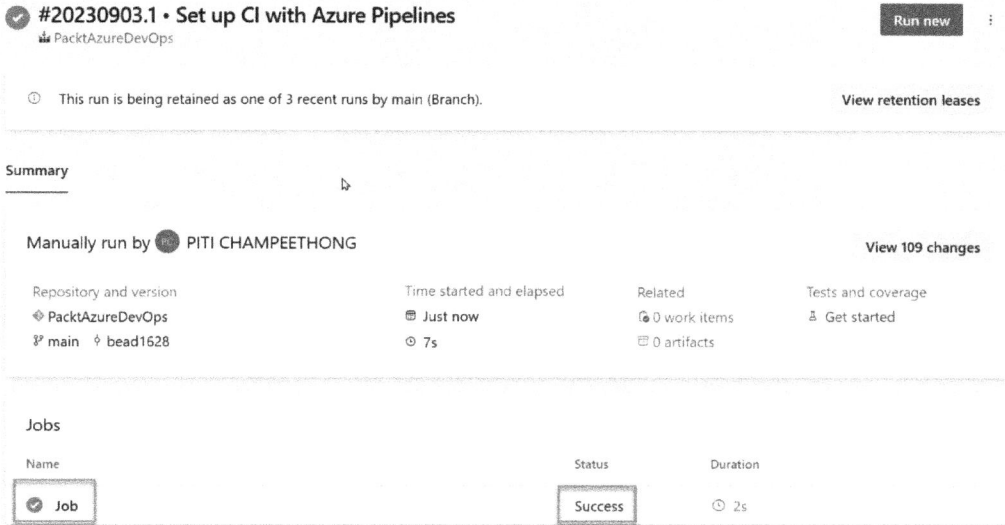

Figure 4.9 – A build pipeline result

9. You can edit the pipeline by clicking on the ellipses (**…**) next to the **Run new** button shown in the following screenshot and then **Edit pipeline**:

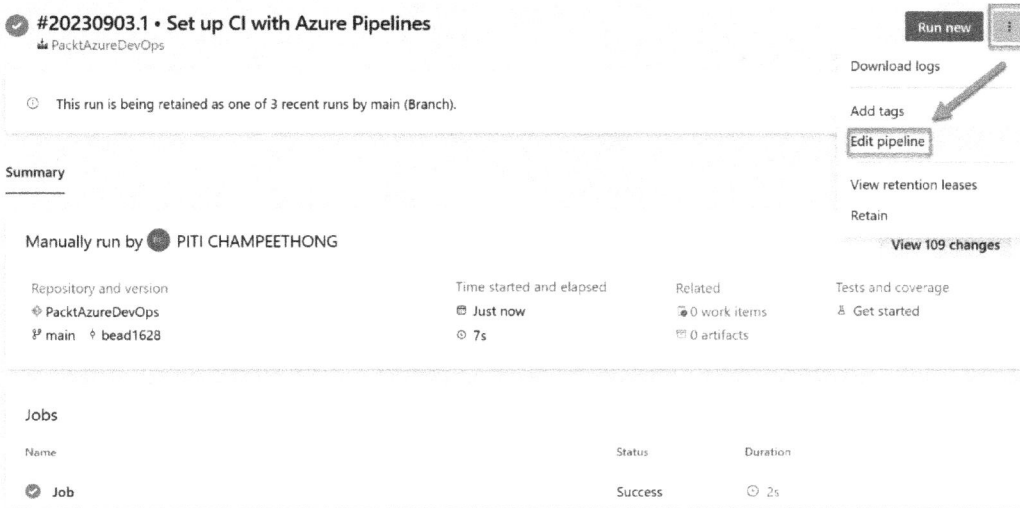

Figure 4.10 – Edit pipeline

10. You can now view the starter of the YAML file structure. Let's look at the structure of the example YAML and describe each part:

   A.   The **main** branch of Azure Repos keeps the YAML file.

   B.   This is a repository name.

   C.   This is a YAML filename.

   D.   A build pipeline will run on any changes in the main branch.

   E.   A build pipeline will run on the Ubuntu operating system.

   F.   A script task contains a single line.

   G.   A script task contains multiple lines.

   These components are shown in the following screenshot:

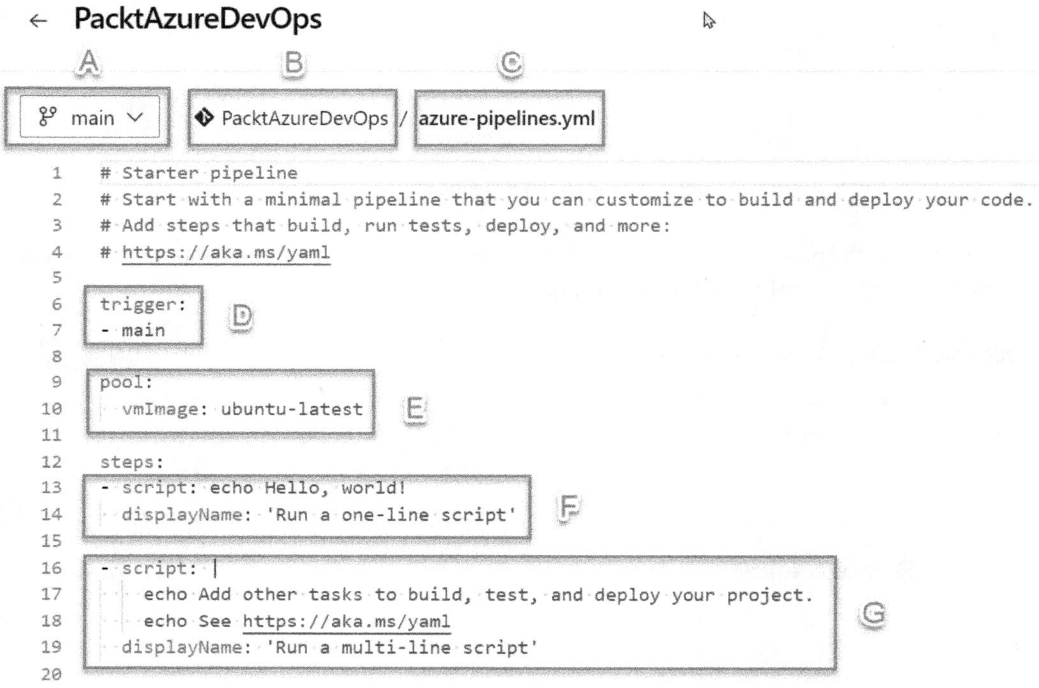

Figure 4.11 – YAML file structure

11. You can see the result after running the pipeline based on the YAML file in the previous screenshot by clicking on a job:

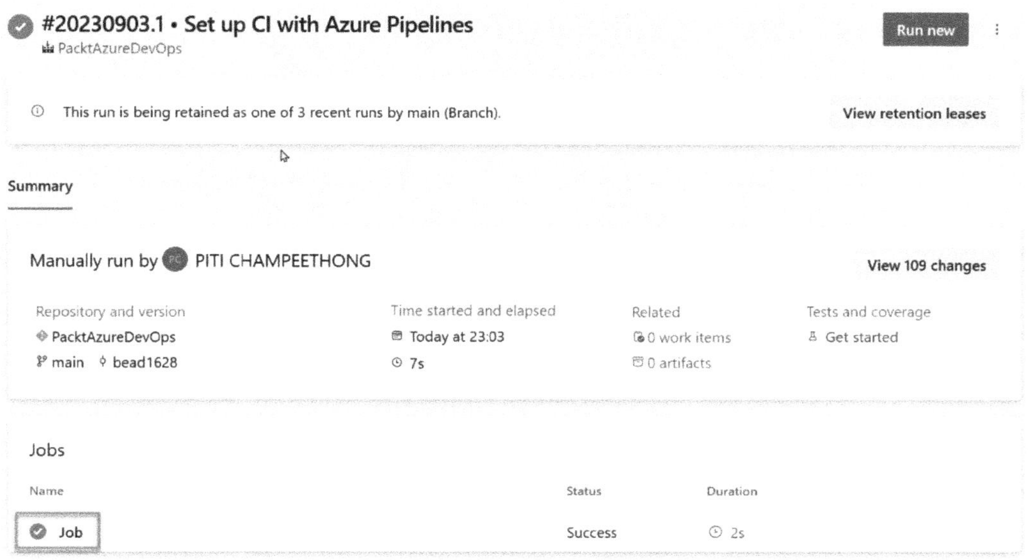

Figure 4.12 – Display job result details

12. Click on **Run a one-line script**, which will display the **Hello, world!** Text. This is an example of when you would like to display the message inside the task of the Azure pipeline:

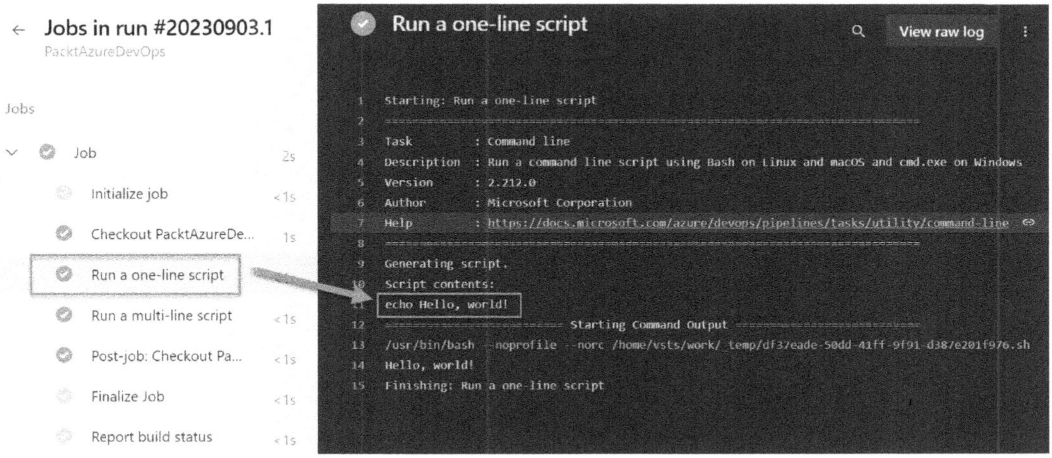

Figure 4.13 – Job steps with results

You learned how to create a simple build pipeline using a YAML file for this section. Azure Repos keeps the YAML file, and you can view its history with the help of Git. In the next section, you will learn how to create a release pipeline using a YAML file that contains stages, jobs, and tasks.

Creating a release pipeline using YAML

This section will teach you how to create a release pipeline using YAML. You will also learn how to create stages, jobs, and tasks in YAML format. To do this, follow these steps:

1. Edit the existing pipeline by clicking on the ellipses (**...**) next to it and then click on **Edit pipeline**:

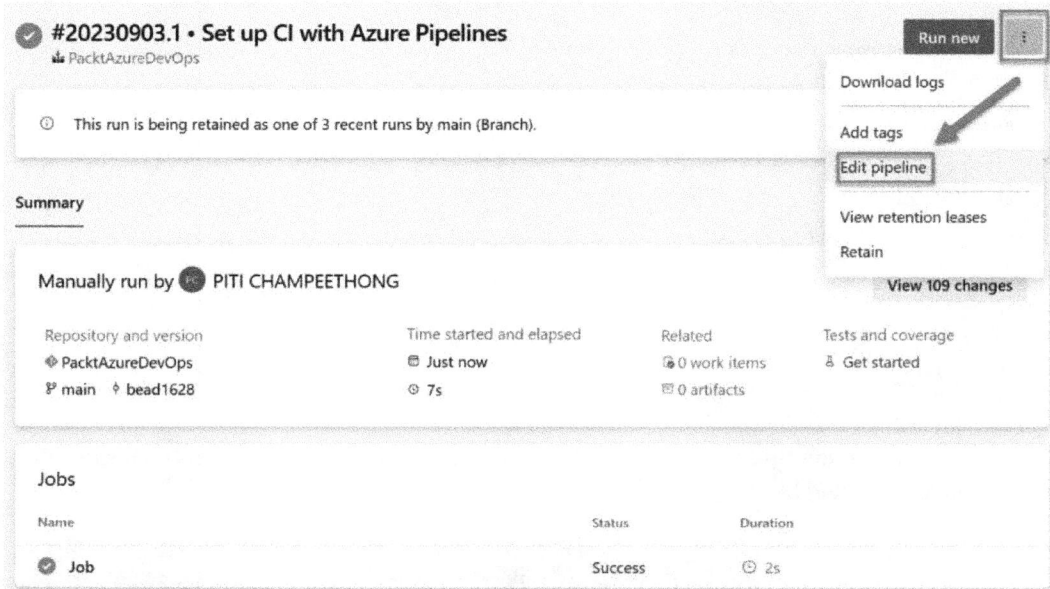

Figure 4.14 – Editing an existing pipeline

2. Replace all contents of the existing `azure-pipelines.yml` file, as shown in the following screenshot:

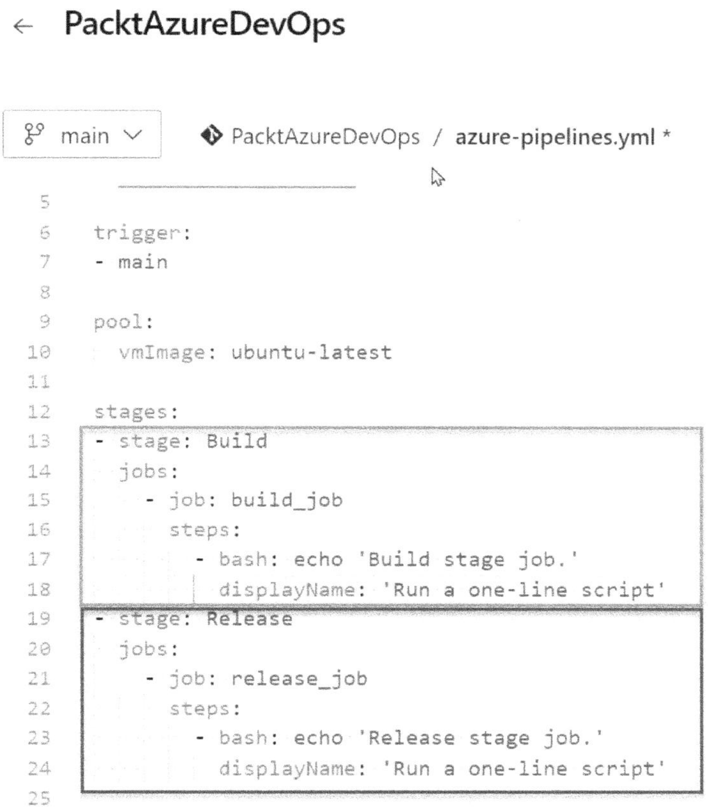

← **PacktAzureDevOps**

```
 ⑧ main ∨        ◆ PacktAzureDevOps / azure-pipelines.yml *
                                          ⤡
  5
  6    trigger:
  7    - main
  8
  9    pool:
 10      vmImage: ubuntu-latest
 11
 12    stages:
 13    - stage: Build
 14      jobs:
 15        - job: build_job
 16          steps:
 17            - bash: echo 'Build stage job.'
 18              displayName: 'Run a one-line script'
 19    - stage: Release
 20      jobs:
 21        - job: release_job
 22          steps:
 23            - bash: echo 'Release stage job.'
 24              displayName: 'Run a one-line script'
 25
```

Figure 4.15 – Advanced pipeline with two stages

There are two stages, as shown in the preceding screenshot:

- The first stage will display Build stage job

- The second stage will display Release stage

This scenario displays build and release stages suitable for building the application (the Build stage) and deploying the application (the Release stage). It is easy to find errors or issues because if the build stage fails, it will not continue to run the release stage.

3. You can validate the syntax of the YAML file by clicking on **...** next to **Save** and clicking **Validate**:

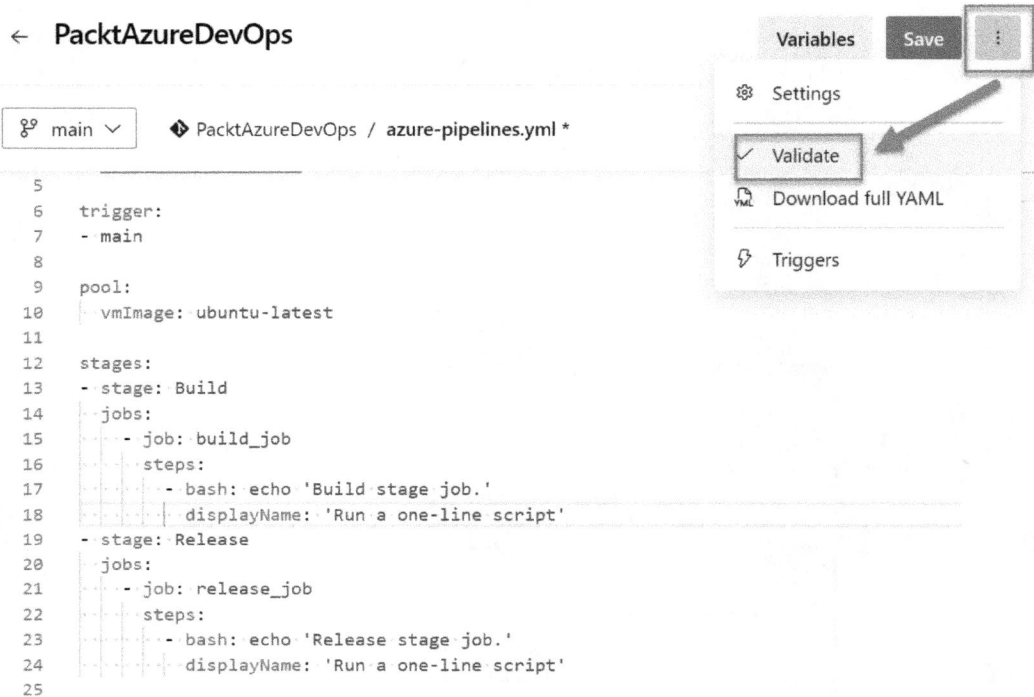

← **PacktAzureDevOps**

	Variables	Save	⋮

⚙ Settings

✓ Validate

🗋 Download full YAML

⚡ Triggers

```
    ⌥ main ∨        ◈ PacktAzureDevOps / azure-pipelines.yml *

 5
 6    trigger:
 7    - main
 8
 9    pool:
10      vmImage: ubuntu-latest
11
12    stages:
13    - stage: Build
14      jobs:
15        - job: build_job
16          steps:
17            - bash: echo 'Build stage job.'
18              displayName: 'Run a one-line script'
19    - stage: Release
20      jobs:
21        - job: release_job
22          steps:
23            - bash: echo 'Release stage job.'
24              displayName: 'Run a one-line script'
25
```

Figure 4.16 – Validate the YAML file

4. You will see the following message if the YAML file is valid:

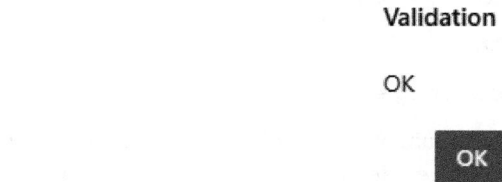

Validation

OK

OK

Figure 4.17 – Valid YAML file

If the YAML file is invalid, you will see the following error message explaining which line has a problem:

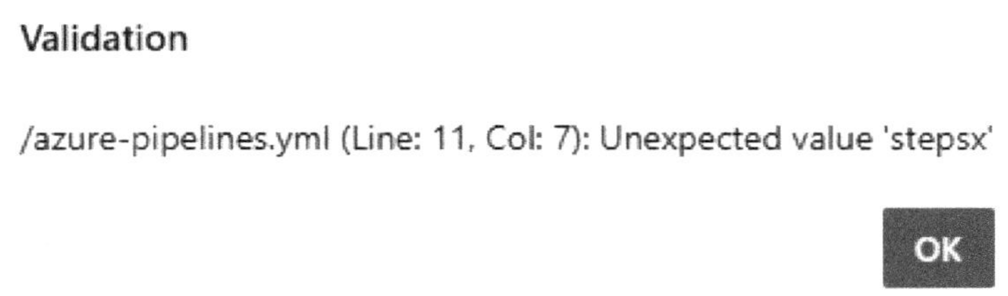

Validation

/azure-pipelines.yml (Line: 11, Col: 7): Unexpected value 'stepsx'

OK

Figure 4.18 – Invalid YAML file

5. Click **Save | Run pipeline**. You can see the result of a pipeline contains two stages:

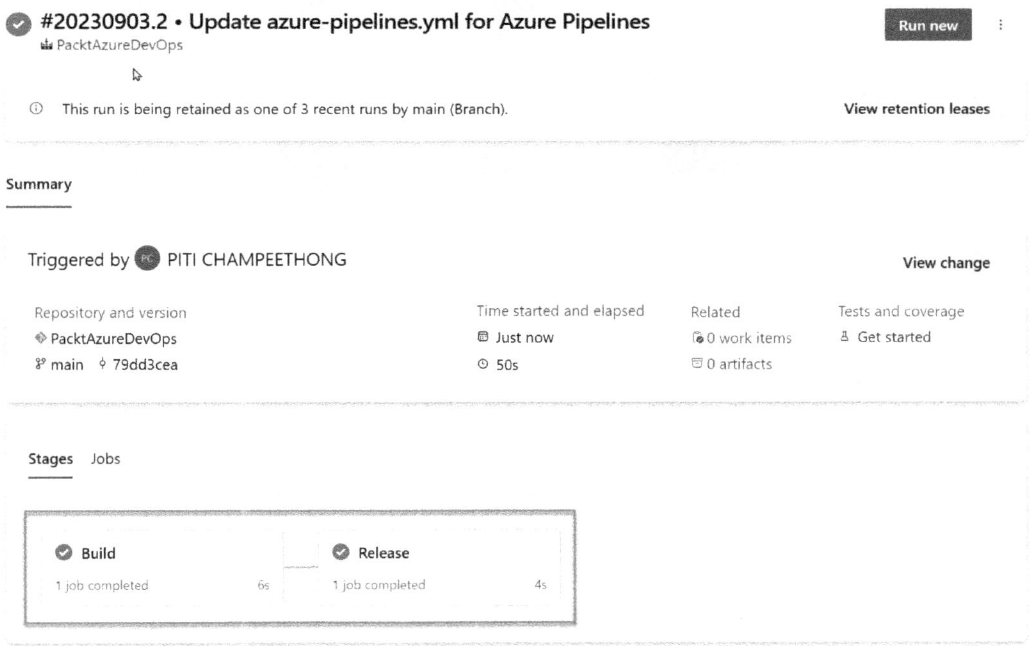

Figure 4.19 – Stage details

6.   You can rerun a specific stage by expanding it and clicking on **Rerun stage**:

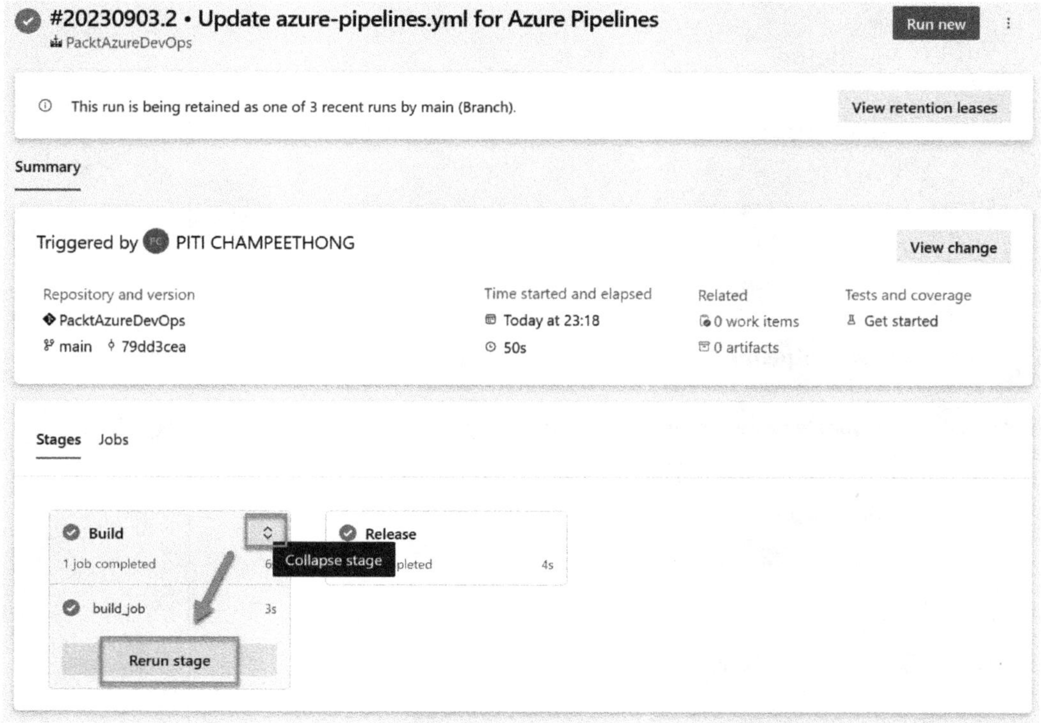

Figure 4.20 – Rerun stage

You have learned how to create a release pipeline with two stages containing a job. You have also learned about the benefit of using stages because you can rerun the stage you have a problem with and don't need to rerun the whole pipeline. The following section will teach you how to clone, export, and import YAML to create a new pipeline.

Cloning, exporting, and importing a YAML pipeline

This section will teach you how to clone, export, and import a YAML pipeline from the Azure DevOps portal. These actions will help you save time when you need to duplicate the same template and adjust it. If you need to make a new Azure pipeline, you can do so by cloning it from an existing one. Let's look at the steps that are to be followed to perform these tasks:

- **Clone**: You can clone a pipeline using copy and paste, which is the easy, fast way to clone a pipeline.

- **Export and import**: The following steps show you can export an entire YAML file from a pipeline:

  A. You can export a YAML pipeline by clicking on **Edit**:

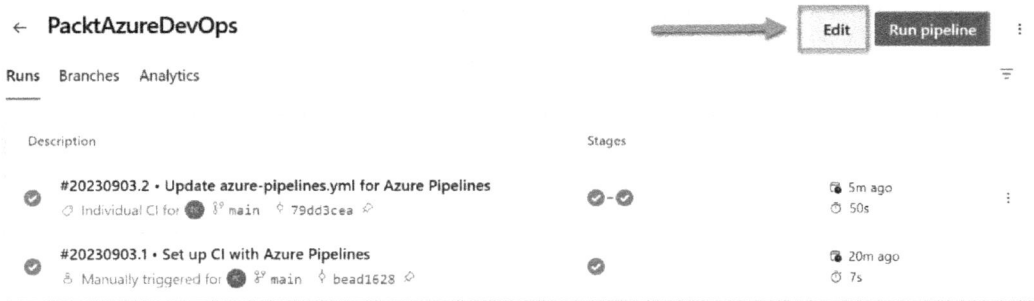

Figure 4.21 – Edit pipeline

  B. Click on **…** | **Download full YAML** to download a file:

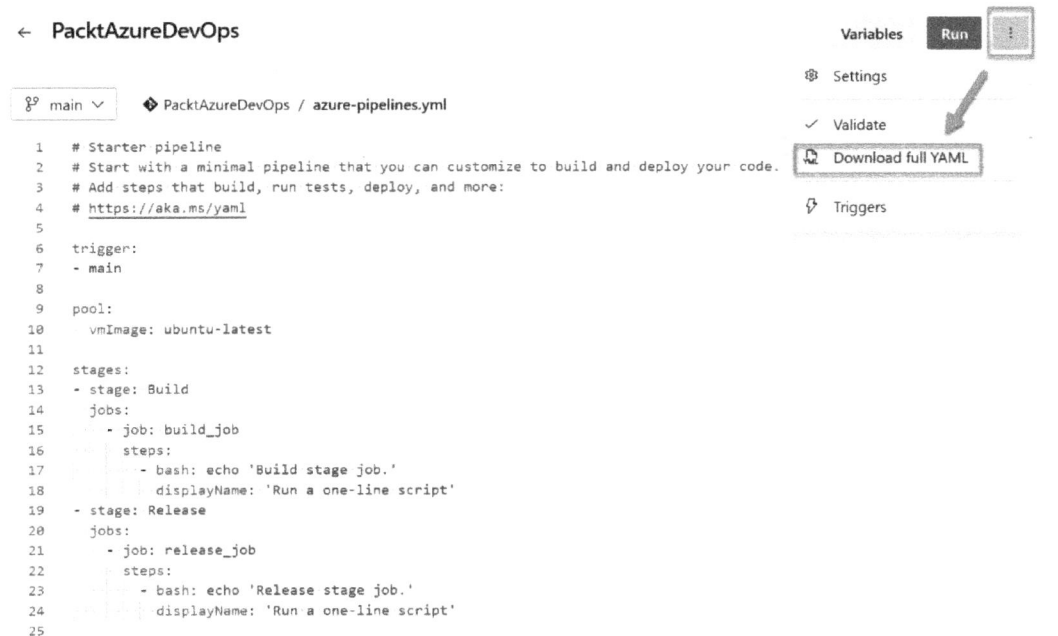

Figure 4.22 – Download the complete YAML

  C. Open a downloaded file and copy and paste it into the new pipeline you created.

In the section, you learned how easy it is to download, copy, and paste a YAML file into the new pipeline. Now let's look at some advantages and drawbacks of using YAML-based pipelines.

Complex YAML configurations

The YAML syntax supports several complex configurations that allow for the modularization and reuse of YAML files. Some examples of these are template reuse and the implementation of template expressions. We'll explore how these features work in the following sections.

YAML template reuse

When working with large projects and several applications being developed by the same members of a team, it is helpful to define common templates to reuse instead of writing everything from scratch for the CI/CD needs of every application. For this purpose, Azure Pipelines supports making references to templates to reuse steps, jobs, and stages. This is particularly helpful in reducing the duplication of YAML because all applications or deployment processes are the same within the project. It is also possible to include parameters in the templates, to pass values that can be used in the referenced template to customize the behavior.

Let's look at the following scenario where an Azure pipeline builds the same application twice with two different build configurations in the .NET language. The file shown in the following screenshot defines one parameter, buildConfiguration, and three steps to perform the installation of the NuGet tool (NuGetToolInstaller@1), a NuGet restore of dependencies (NuGetCommand@2), and building the solution with the VS build utility (VSBuild@1):

```
 1  # templates/dotnet-build-steps.yml
 2  parameters:
 3  - name: buildConfiguration
 4    type: string
 5    default: debug
 6    values:
 7    - debug
 8    - release
 9
10  - task: NuGetToolInstaller@1
11  - task: NuGetCommand@2
12    inputs:
13      restoreSolution: '**\*.sln'
14  - task: VSBuild@1
15    inputs:
16      solution: '**\*.sln'
17      msbuildArgs: '/p:DeployOnBuild=true /p:WebPublishMethod=Package /p:PackageAsSingleFile=true'
18      platform: 'any cpu'
19      configuration: {{ parameters.buildConfiguration }}
```

Figure 4.23 – dotnet-build-steps template with parameters

You can see in the following example how the same `dotnet-build-steps.yml` file can now be used to build the same application in two different agents, `linux-latest` and `windows-latest`, with the same exact steps and having the possibility to add other tasks before and after:

```
1  # File: azure-pipelines.yml
2
3  jobs:
4  - job: Build-Debug
5    pool:
6      vmImage: 'linux-latest'
7      steps:
8      - template: templates/dotnet-build-steps.yml  # Template reference
9        parameters:
10         buildConfiguration: 'debug'
11
12 - job: Build-Release
13   pool:
14     vmImage: 'windows-latest'
15     steps:
16     - script: echo This script runs before the template's steps, only on Windows.
17     - template: templates/dotnet-build-steps.yml  # Template reference
18       parameters:
19         buildConfiguration: 'release'
20     - script: echo This step runs after the template's steps.
```

Figure 4.24 – Azure pipeline with template references

This feature allows for great flexibility, reducing duplicate code in your YAML pipelines, and standardization across pipelines, which helps reduce the chances of errors.

More complex configurations can also include having all templates in a separate repository, where another team is responsible for putting together these building blocks to help the teams in charge of CI/CD pipelines, as shown in the following screenshot:

```
1  # Repo: Pack/Product
2  # File: azure-pipelines.yml
3  resources:
4    repositories:
5      - repository: templates
6        type: git
7        name: Packt/Templates
8
9  jobs:
10 - template: common.yml@templates  # Template reference
```

Figure 4.25 – Template reference from another repository

Let's look now at how we can use template expressions next.

YAML template expressions

Expressions are a custom syntax capability in Azure Pipelines that allows you to dynamically resolve values during runtime. Think of this as control logic in the execution of your templates. There are too many expression types to cover in this chapter, but it is important to know that it is possible to do the following:

- Evaluate literals and variables
- Use built-in functions such as `coalesce`, `contains`, `eq`, `format`, and many others to evaluate logical conditions or transform values
- Use built-in functions to evaluate the job status
- Use conditions to conditionally insert variable values or tasks
- Loop through parameters with the `each` keyword
- Evaluate dependencies on previous jobs or stages, such as status or output variables

The following screenshot shows an example of how expressions can support performing the building and testing of an application using two different sets of tools (the `msbuild` or `dotnet` CLI tools) and picking the tool based on a parameter:

```
 1  # File: steps/build.yml
 2
 3  parameters:
 4  - name: 'toolset'
 5    default: msbuild
 6    type: string
 7    values:
 8    - msbuild
 9    - dotnet
10
11  steps:
12  # msbuild
13  - ${{ if eq(parameters.toolset, 'msbuild') }}:
14    - task: msbuild@1
15    - task: vstest@2
16
17  # dotnet
18  - ${{ if eq(parameters.toolset, 'dotnet') }}:
19    - task: dotnet@1
20      inputs:
21        command: build
22    - task: dotnet@1
23      inputs:
24        command: test
```

Figure 4.26 – Template expressions example

With expressions, you have full control of how to define your pipelines and dynamically execute steps based on progress, not just simply defining a static set of steps.

Now that you have understood these complex configurations, let's talk about the advantages and limitations of YAML-based pipelines.

Advantages and limitations of YAML-based pipelines

First, let's look at the benefits of using YAML-based pipelines:

- YAML pipelines are stored as code in your version control system, such as Azure Repos. This means they can be versioned, branched, and reviewed like any other code, providing better collaboration and traceability.

- YAML pipelines allow you to introduce changes in a controlled way when implementing a branch strategy that isolates work from different team members. This ensures changes are tested without affecting other members of the team until completed.

- YAML pipelines make reproducing the build and release process consistently across different environments easy. This helps reduce configuration drift and ensures consistent results.

- You can control your build and release processes. You can define the steps, dependencies, and conditions, enabling you to customize pipelines to meet your specific needs.

Despite these benefits, using these pipelines may have drawbacks such as the following:

- YAML-based pipelines require knowledge of the YAML language and Azure Pipelines syntax, which can be a learning curve for everyone unfamiliar with Azure Pipelines

- The YAML-based pipelines syntax only works in Azure Pipelines and cannot be migrated as is to a different CI/CD tool

- YAML-based pipelines have a limit of 4 MB for the size of the extended YAML, which could make it difficult for extremely complex CI/CD processes to be defined

- Validating YAML pipelines in release stages beyond the initial developer environment might be difficult or impossible due to the inability to deploy to such environments

Now that you're familiar with YAML-based pipelines, let's wrap up this chapter.

Summary

This chapter taught you how to build and release pipelines using YAML. Using this method is more powerful for developers than using the classic editor on the Azure DevOps portal. Developers can save YAML files in their Azure Repos, which will help them review each revision's pipeline. With YAML-based pipelines, you can deliver more efficient, transparent, and developer-centric CI/CD processes. You also learned about complex scenarios and how to reduce YAML, reuse templates, and add dynamic behavior through expressions. Finally, you learned about the pros and cons of YAML-based pipelines.

In the next chapter, you will learn in depth about implementing a build and release pipeline using YAML and how to reuse build tasks using Node.js, NPM, .NET, and Docker to build a pipeline.

Part 2:
Azure Pipelines in Action

Now that we have learned the basics, it's time to learn how to use our pipelines for build and deployment purposes, from applications to automated provisioning and configuration of infrastructure, including the testing and security tools involved in these processes.

This part has the following chapters:

- *Chapter 5, Implementing the Build Pipeline Using Deployment Tasks*
- *Chapter 6, Integrating Testing, Security Tasks, and Other Tools*
- *Chapter 7, Monitoring Azure Pipelines*
- *Chapter 8, Provisioning Infrastructure Using Infrastructure as Code*

5

Implementing the Build Pipeline Using Deployment Tasks

In the previous chapter, we created a pipeline using YAML, and we learned the process of creating jobs and tasks in the YAML format and exporting and importing a build pipeline. This chapter will dive deep into creating a pipeline using standard tasks. By the end of this chapter, you will have learned how to create a build pipeline for web application development, including Node.js, .NET Core, Docker, and Microsoft SQL Server, both on-premises and in Azure, using beginner-friendly tasks that make it easy to understand the concept.

We will cover the following topics:

- Working with Node.js and **Node Package Manager** (**NPM**) tasks
- Working with .NET Core CLI tasks
- Working with Docker tasks
- Working with SQL Server deployment tasks

Let's start by learning how to create a pipeline using Node.js and NPM tasks.

Working with Node.js and NPM tasks

You need to use Node.js and NPM commands to build and deploy Node.js applications. There are many predefined tasks to build such applications in an Azure pipeline. Follow these steps to create a pipeline using Node.js and NPM tasks:

1. After logging in to the Azure DevOps portal, select your organization, navigate to the **Pipelines** page, and click on **New pipeline**:

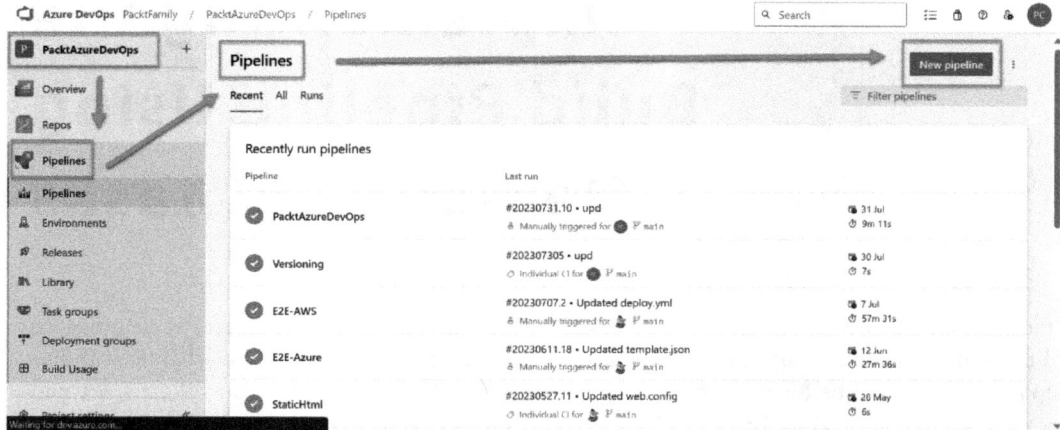

Figure 5.1 – A new pipeline

2. Select **Azure Repos Git**, which is a source code repository for this demo:

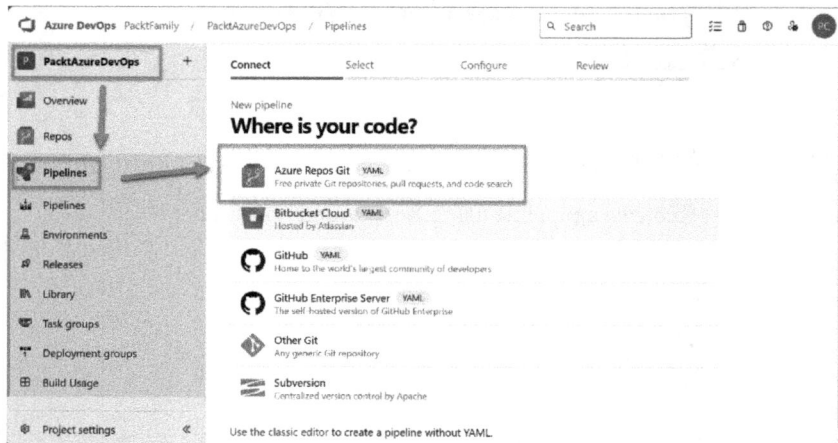

Figure 5.2 – Selecting Azure Repos Git

3.   Select the **PacktAzureDevOps** repository that we created in *Chapter 2*:

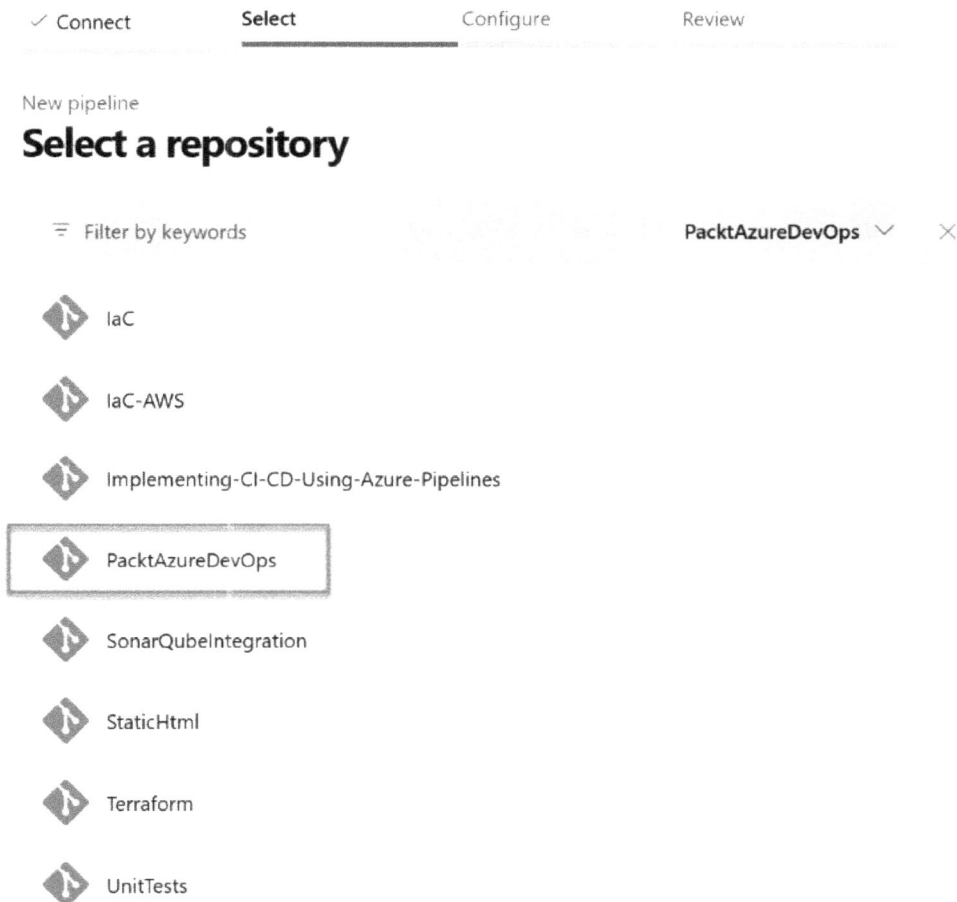

✓ Connect	**Select**	Configure	Review

New pipeline
Select a repository

≡ Filter by keywords   **PacktAzureDevOps** ∨   ✕

◆   IaC

◆   IaC-AWS

◆   Implementing-CI-CD-Using-Azure-Pipelines

◆   PacktAzureDevOps

◆   SonarQubeIntegration

◆   StaticHtml

◆   Terraform

◆   UnitTests

Figure 5.3 – Selecting a repository

4. Click on **Show more**:

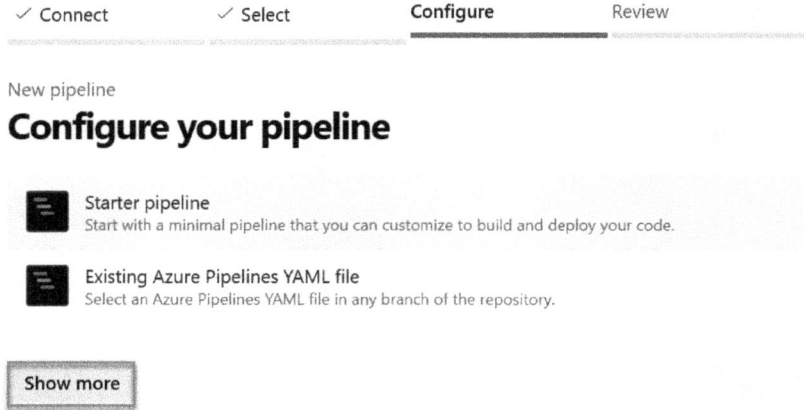

Figure 5.4 – Showing more tasks

5. Select the **Node.js** option:

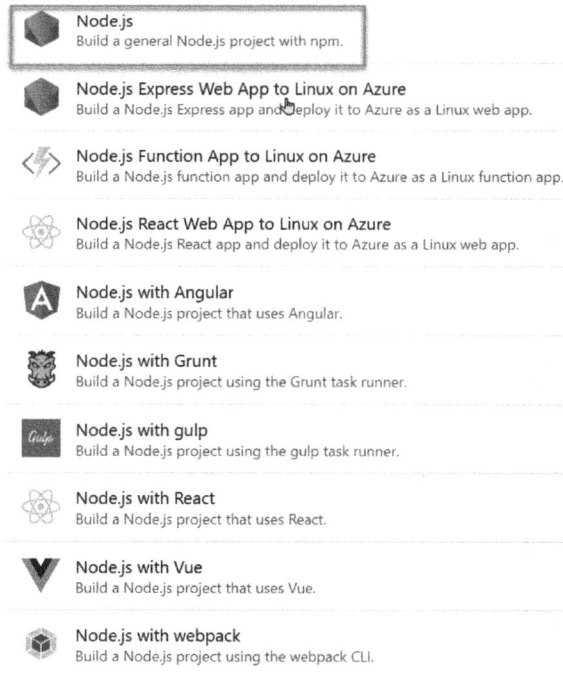

Figure 5.5 – Selecting Node.js

6. You can rename the default filename, which is `azure-pipelines-1.yml`, by clicking on it and changing it to `node.yml`:

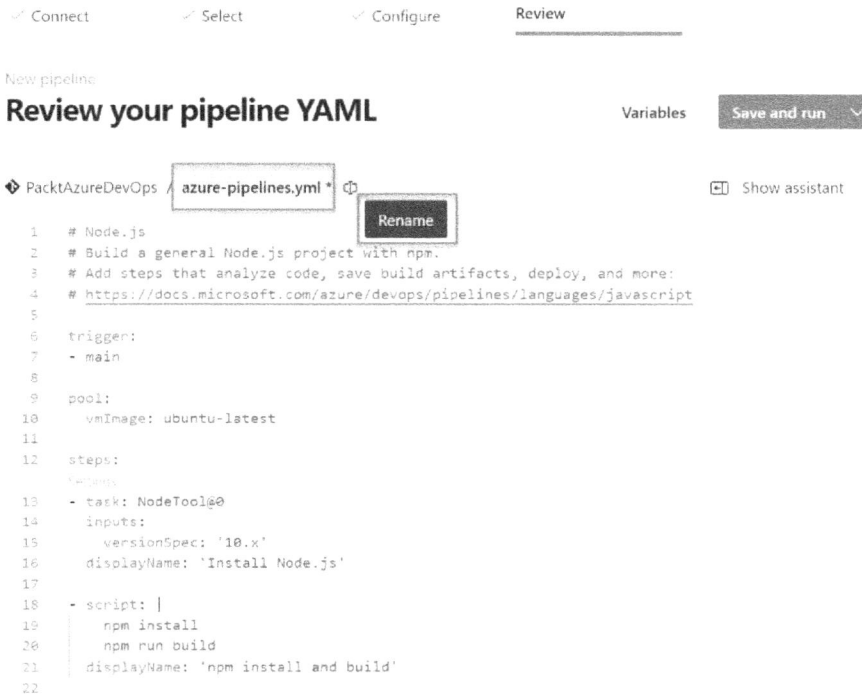

Figure 5.6 – Editing the name of the file

7. Click on **Save and run | Save**:

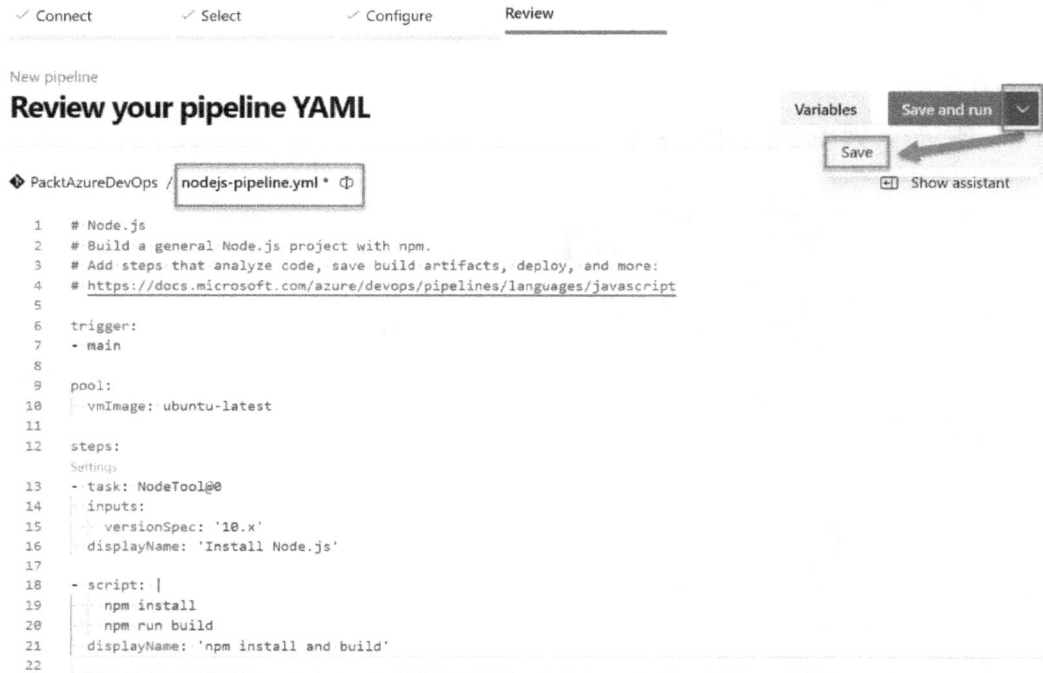

Figure 5.7 – Saving a pipeline file

After you select a template of Node.js and NPM tasks, you can continue changing the default NPM command that matches your Azure pipeline, such as which version of Node.js you need. The following section will show you how to create tasks for .NET Core.

Working with .NET Core CLI tasks

For .NET applications, you must use .NET Core CLI commands to build and deploy .NET applications. There are many predefined tasks to build .NET applications in an Azure pipeline. Follow these steps to create a pipeline using .NET Core CLI tasks:

1. Follow *steps 1 to 3* from the previous section for Node.js and NPM tasks.

2. Select **Starter pipeline**:

✓ Connect   ✓ Select   **Configure**   Review

New pipeline

Configure your pipeline

Starter pipeline
Start with a minimal pipeline that you can customize to build and deploy your code.

Existing Azure Pipelines YAML file
Select an Azure Pipelines YAML file in any branch of the repository.

Show more

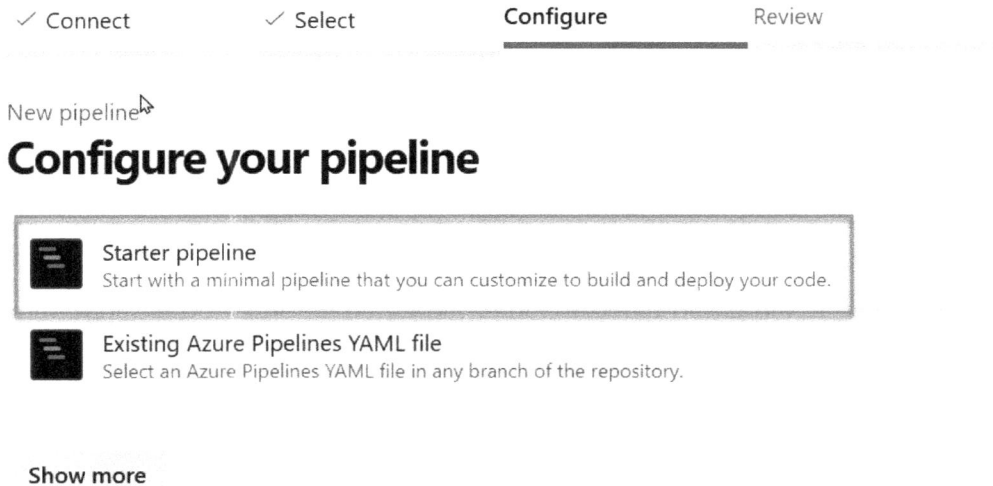

Figure 5.8 – Selecting the Starter pipeline option

3. Rename the file from the default name to make it easier to understand what the YAML file is for:

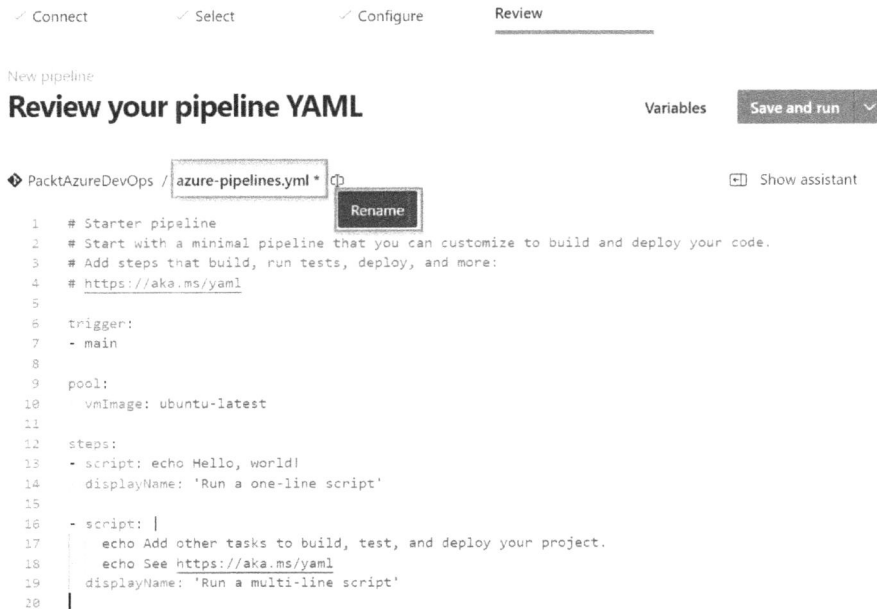

✓ Connect   ✓ Select   ✓ Configure   Review

New pipeline

Review your pipeline YAML   Variables   Save and run ⌄

◆ PacktAzureDevOps / azure-pipelines.yml *   Show assistant

                                        Rename

```
1    # Starter pipeline
2    # Start with a minimal pipeline that you can customize to build and deploy your code.
3    # Add steps that build, run tests, deploy, and more:
4    # https://aka.ms/yaml
5
6    trigger:
7    - main
8
9    pool:
10     vmImage: ubuntu-latest
11
12   steps:
13   - script: echo Hello, world!
14     displayName: 'Run a one-line script'
15
16   - script: |
17       echo Add other tasks to build, test, and deploy your project.
18       echo See https://aka.ms/yaml
19     displayName: 'Run a multi-line script'
20
```

Figure 5.9 – Renaming a pipeline file

4. Select the **Use .NET Core** task and click **Add**:

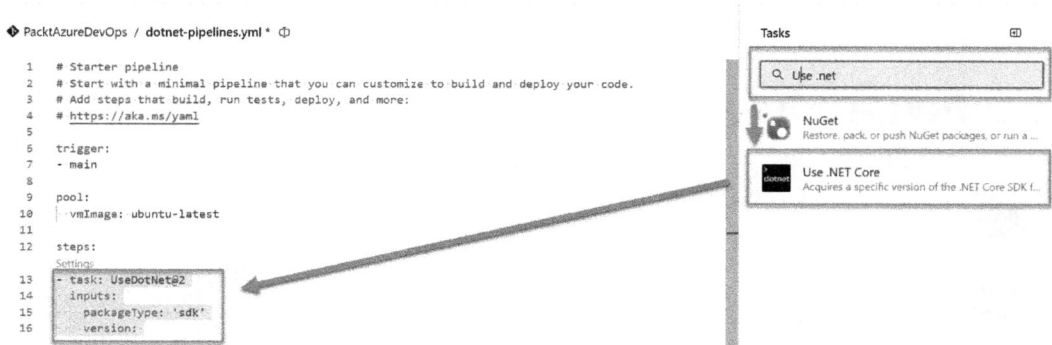

Figure 5.10 – Selecting the Use .NET Core task

5. Update the `version` property to use .NET 6:

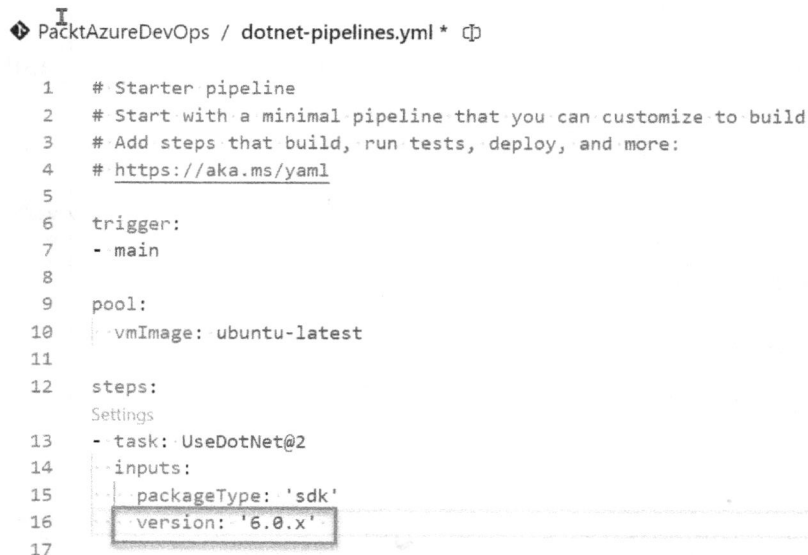

Figure 5.11 – Updating the .NET version

6. Select the **.NET Core** task and click on **Add**:

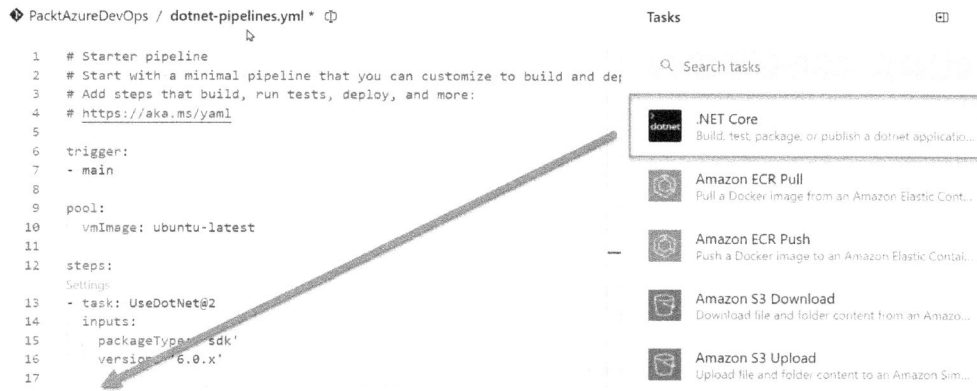

Figure 5.12 – Selecting the .NET Core task

7.   Review the two predefined .NET tasks:

- `UseDotNet@2` is used to install the .NET compiler version 6.0.x

- `DotNetCoreCLI@2` is the .NET command to run a specific command, which is the `build` command

These are shown in the following screenshot:

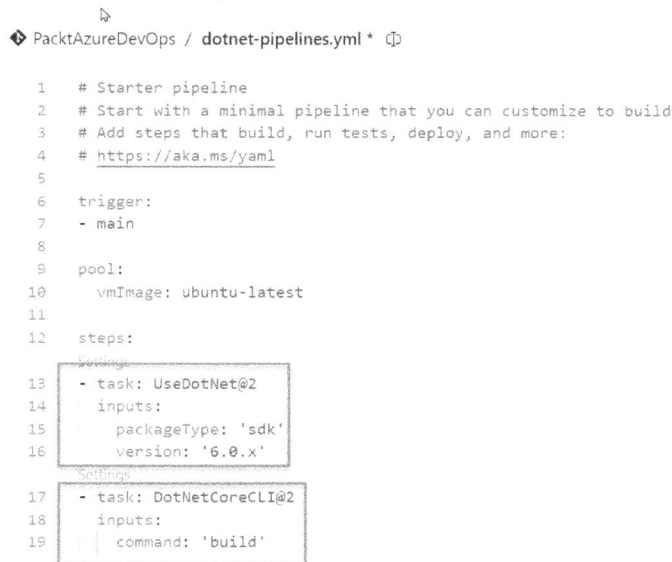

Figure 5.13 – A view of the .NET Core build task

After you create a starter task for the .NET CLI command, you can continue to customize your tasks by using the `DotNetCoreCLI@2` command, which specifies the `build` command to be used to build the .NET application, from source code to the .NET binary files. The following section will show you how to work with Docker tasks for containerized applications.

Working with Docker tasks

For cloud-native applications, you need to use Docker commands to build and deploy cloud-native applications. There are many predefined tasks for building cloud-native applications in an Azure pipeline. You can perform the following steps to create a pipeline using Docker tasks:

1. Follow *steps 1 to 4* as described in the previous section for .NET Core CLI tasks.

2. Rename the file for the Docker pipeline:

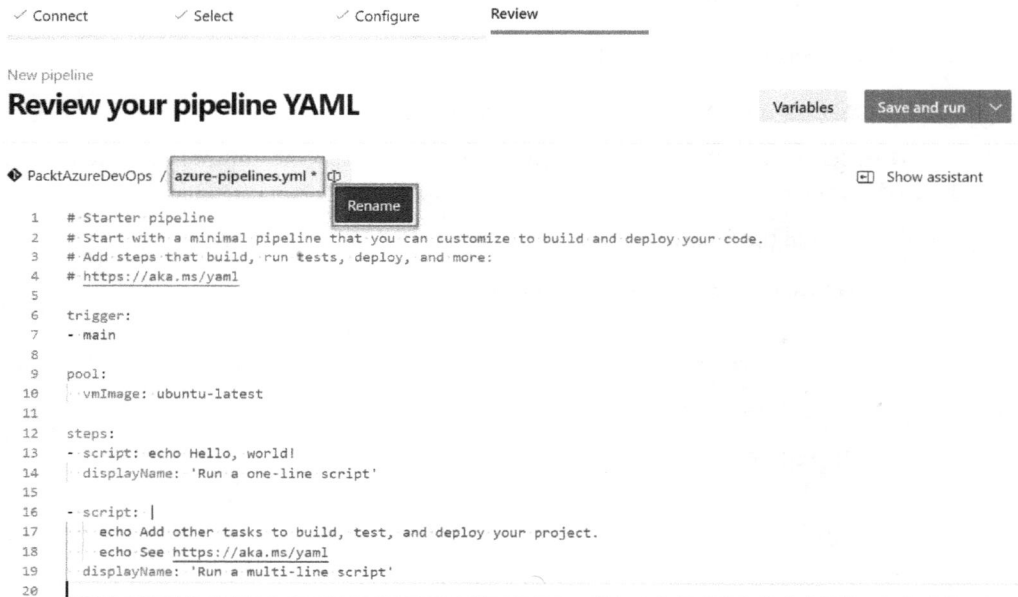

Figure 5.14 – Renaming the file

3. Select the **Docker CLI installer** task, click on **Add**, and update `DockerInstaller@0`:

```
- task: DockerInstaller@0
  inputs:
    dockerVersion: '17.09.0-ce'
```

The following screenshot shows you how to add a **Docker CLI installer** task and fill in the details on this task:

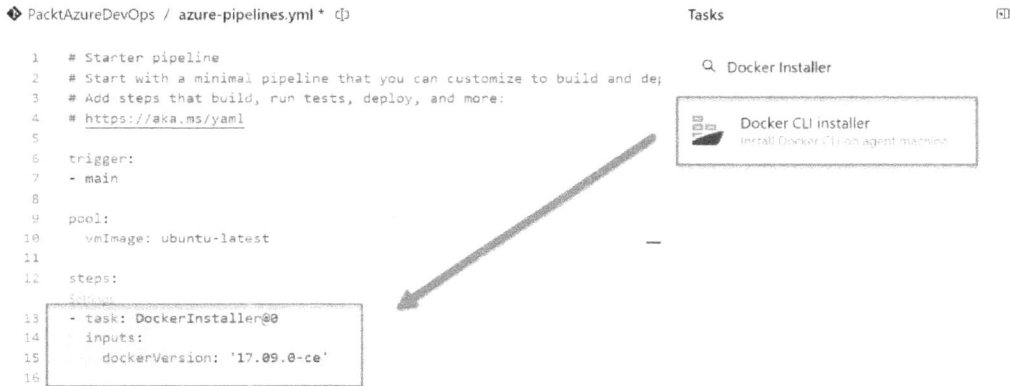

Figure 5.15 – Adding a Docker CLI installer task

4. Select the **Docker** task and click on **Add**, and you will see the following code. This is the Docker task to build and push images in one task:

```
- task: Docker@2

  inputs:
    command: 'buildAndPush'
    Dockerfile: '**/Dockerfile'
```

The following screenshot shows you how to add a Docker task and fill in the details:

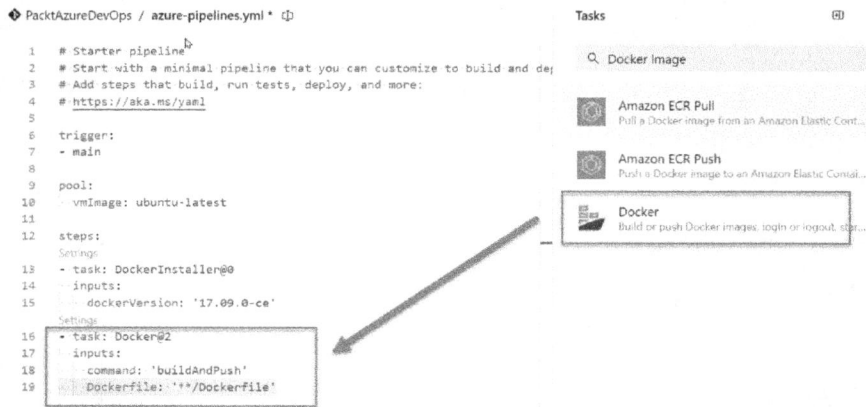

Figure 5.16 – Adding a Docker task

For easy error handling, you can replace the buildAndPush task with separate *build* and *push* tasks. In the *push* task, you set the condition value as succeeded(), which ensures that the task will only run if the previous steps (in this case, the build task) are completed successfully.

You can use the following code to replace the buildAndPush task:

```
- task: Docker@2
  displayName: 'Build Docker image'
  inputs:
    command: build
    Dockerfile: '**/Dockerfile'
    tags: latest

- task: Docker@2
  displayName: 'Push Docker image'
  inputs:
    command: push
  condition: succeeded()
```

The difference is that the buildAndPush task will build and push the image in one task, which means if you need to add a task between building and pushing, you cannot do that.

5. Next, select the **Command line** task and click **Add**. Update the command line, as shown in the following code snippet:

```
- task: CmdLine@2
  inputs:
    script: |
        docker login <docker hub url> -u <your username> -p <your
password>
        docker push <your repository>:<your tag>
```

The following screenshot shows you how to add a **Command line** task and fill in the details:

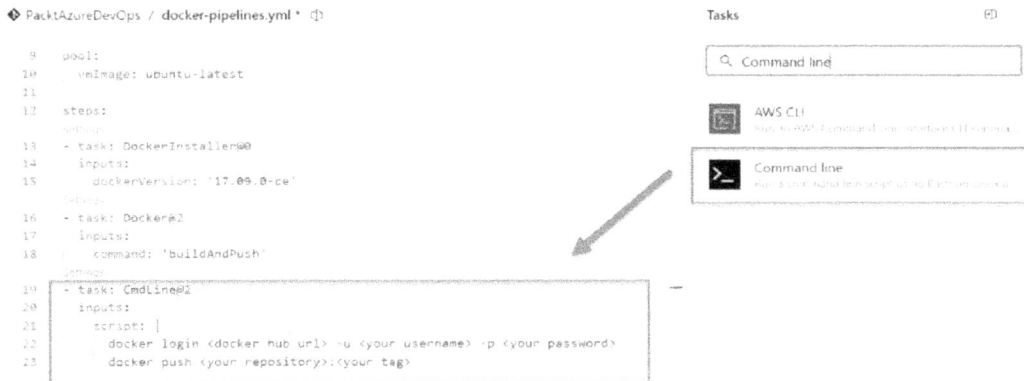

Figure 5.17 – Adding a Command line task

After running this pipeline, you will see the Docker image on your Docker Hub.

Working with SQL Server deployment tasks

You need to use SQL Server commands to build and deploy SQL Server applications. There are many tasks that need to be completed to build SQL Server applications in Azure Pipelines; follow these steps to create a pipeline using SQL Server deployment tasks:

1. You can follow *steps 1 to 4* as described in the *Working with .NET Core CLI tasks* section.

2. Rename the file as shown in the following screenshot:

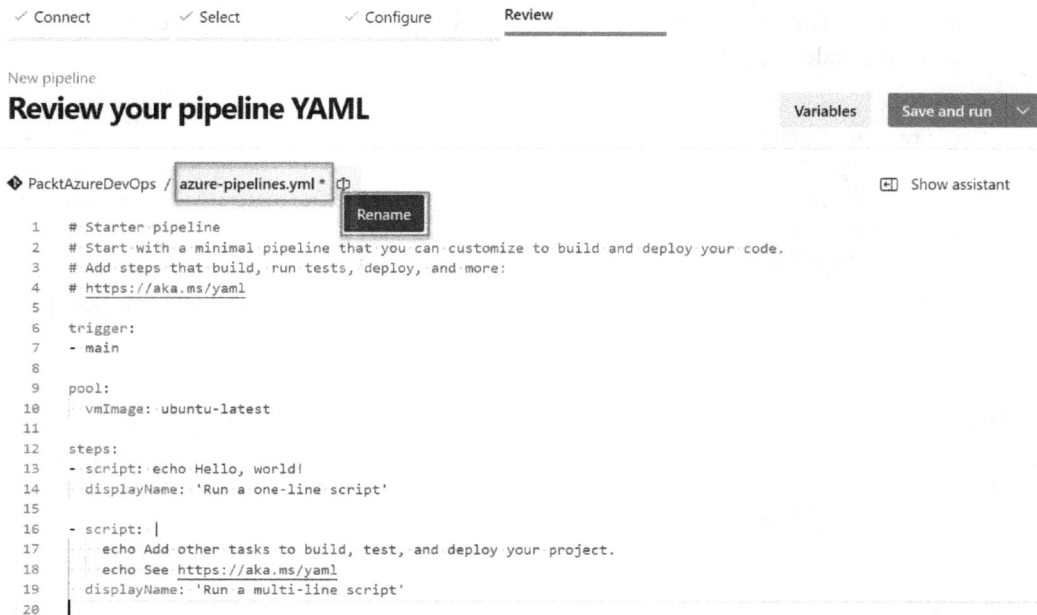

Figure 5.18 – Renaming the file

3. Search for SQL Server database, select the **SQL Server database deploy** task, and enter the following:

```
- task: SqlDacpacDeploymentOnMachineGroup@0
  inputs:
    TaskType: 'sqlQuery'
    SqlFile: 'migrate.sql'
    ExecuteInTransaction: true
    ServerName: 'localhost'
    DatabaseName: 'your_database'
    AuthScheme: 'sqlServerAuthentication'
    SqlUsername: 'your_username'
    SqlPassword: 'your_password'
```

Let's look at each property in detail:

- TaskType: This could be dacpac, sqlQuery, or sqlInline:

  - dacpac means this task will execute the SQL commands from the dacpac file

  - sqlQuery means this task will execute the SQL commands, such as the SELECT, UPDATE, INSERT, and DELETE commands, in a SQL file

- sqlInline means this task will execute the SQL directly as a NOT value in the SQL file

- SqlFile: This is the full name path for a SQL file

- ExecuteInTransaction: If this is set to true, the task will execute a SQL file under the transaction scope

- ServerName: This can be a database server name or an IP

- DatabaseName: This is a database name for execution

- AuthScheme: This could be sqlServerAuthentication, which will use authentication from SQL Server, or windowsAuthentication, which will use authentication from Windows

- SqlUsername: This is a user from SQL Server

- SqlPassword: This is a password from SQL Server

The following screenshot shows you how to add the SQL **Data-Tier Application Package** (**DACPAC**) deployment task and fill in the details:

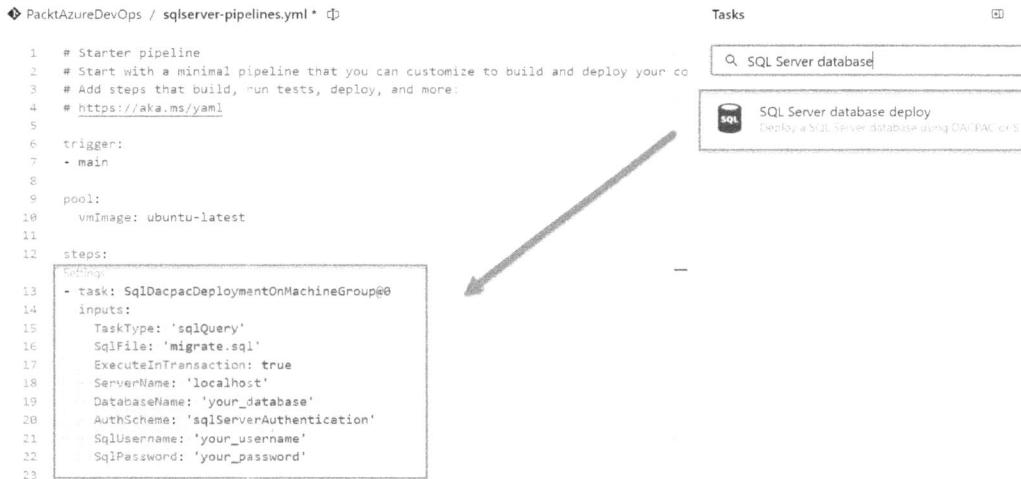

```yaml
# Starter pipeline
# Start with a minimal pipeline that you can customize to build and deploy your co
# Add steps that build, run tests, deploy, and more:
# https://aka.ms/yaml

trigger:
- main

pool:
  vmImage: ubuntu-latest

steps:
- task: SqlDacpacDeploymentOnMachineGroup@0
  inputs:
    TaskType: 'sqlQuery'
    SqlFile: 'migrate.sql'
    ExecuteInTransaction: true
    ServerName: 'localhost'
    DatabaseName: 'your_database'
    AuthScheme: 'sqlServerAuthentication'
    SqlUsername: 'your_username'
    SqlPassword: 'your_password'
```

Figure 5.19 – Adding the SQL Server database deploy task

After you create a SQL Server database deploy task, you can proceed to further customize your tasks. For instance, you can specify the ServerName parameter, which represents the database hostname or the server name of the SQL Server that this task should connect to. After making these customizations, you can save the pipeline file.

Summary

This chapter taught you how to build and release pipelines, using standard NPM, the .NET Core CLI, Docker, and SQL Server deployment tasks. These predefined tasks are popular for building and deploying Node.js and .NET applications. They reduce the time spent by developers creating manual commands when running a pipeline to build those applications.

In the next chapter, you will learn in depth about integrating testing and security tasks to make your code and applications more reliable.

Integrating Testing, Security Tasks, and Other Tools

Now that we have learned the basics of build and release pipelines, it is time to understand how Azure Pipelines can be extended with other tools, to perform additional tasks and be able to include additional capabilities beyond those built into the tool. By the end of this chapter, you will have the skills to go beyond the basics and be able to include tasks to increase the quality of the code produced in builds, detect vulnerabilities before deploying, and use source code from another repository and inclusive artifacts from other locations.

In this chapter, we will cover the following topics:

- Understanding the Azure DevOps extensibility model
- Including automated tests for your build
- Increasing code quality
- Integrating with Jenkins for artifacts and release pipelines

Technical requirements

To complete this chapter, you will need certain extensions. Let's understand the **Azure Devops extensibility model** first and how you can access them. You will find the code for this chapter in the GitHub repository at `https://github.com/PacktPublishing/Implementing-CI-CD-Using-Azure-Pipelines/tree/main/ch06`.

Understanding the Azure DevOps extensibility model

Azure DevOps and its sub-services provide several features that are included by default, but you can customize and extend your experience using extensions that can be developed using standard technologies such as HTML, JavaScript, and CSS.

There is a very flexible model behind all sub-services that you can augment using **extensions** published by individuals and well-known third-party organizations available in the marketplace. You have the option to create your own and publish them as well if you don't find what you need.

The purpose of an extension is to make it easier to encapsulate reusable tasks, use external tools, and even enhance the look and feel of Azure DevOps. For Azure Pipelines, you will find extensions that do the following:

- Make complex and repetitive tasks easier
- Use common **Infrastructure as Code (IaC)** tools easily, such as Terraform or Ansible
- Integrate with SaaS products to improve code quality and security
- Facilitate deployment tasks to cloud providers such as Azure and Amazon Web Services

You can access Visual Studio Marketplace for Azure DevOps at `https://marketplace.visualstudio.com/azuredevops`.

The following screenshot shows what this looks like:

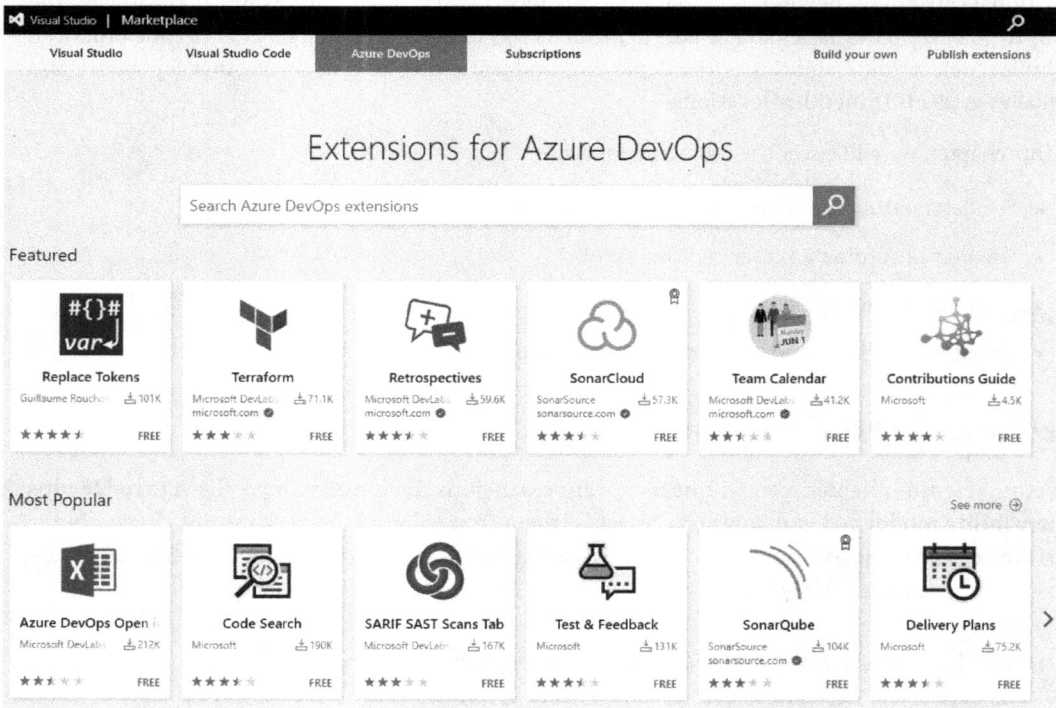

Figure 6.1 – Visual Studio Marketplace for Azure DevOps

Each marketplace extension listing indicates the name of the extension, who published it, whether the publisher is verified, its rating, number of installations, and price. Some are provided for free, while others might require you to pay a fee for them. You can find these extensions by name, category, or tag to make it easier to find what you are looking for.

Installing a code quality assessment tool, SonarQube

Search the marketplace for SonarQube and click on the listing to see its details. Alternatively, go to `https://marketplace.visualstudio.com/items?itemName=SonarSource.sonarqube`.

You should see something similar to the following:

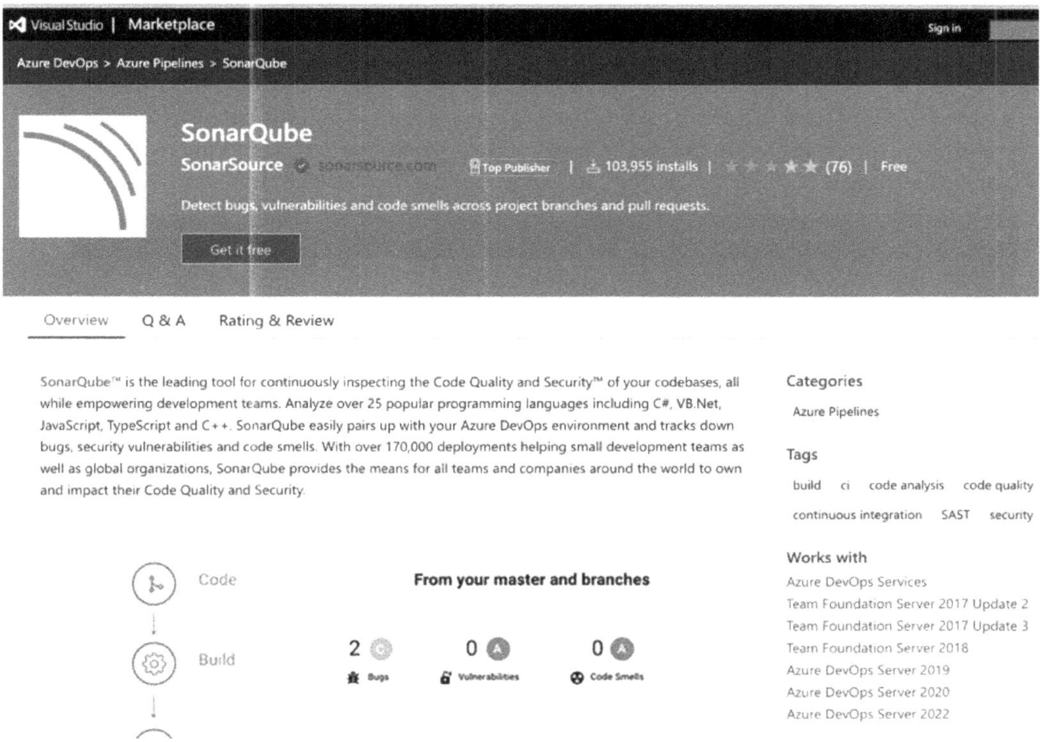

Figure 6.2 – Visual Studio Marketplace listing for the SonarQube extension

Once you have found the extension you want, simply click on the **Get it free** or **Get** button. You will be able to select the Azure DevOps organization to install it in, just in case you have more than one, as shown in the following figure:

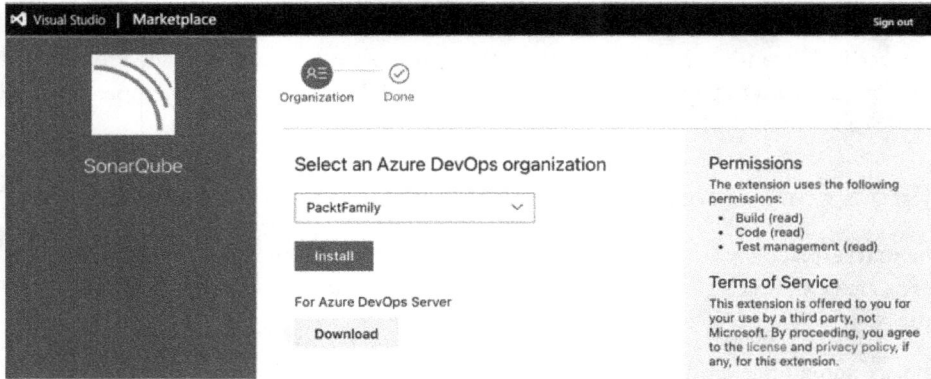

Figure 6.3 – Installing the SonarQube extension from the marketplace

Once you have reviewed the permissions and terms of service and selected the organization to install in, click the **Install** button. It typically takes just a few seconds to install the extension, after which you have the option to proceed to the Azure DevOps organization or go back to the marketplace to find more extensions:

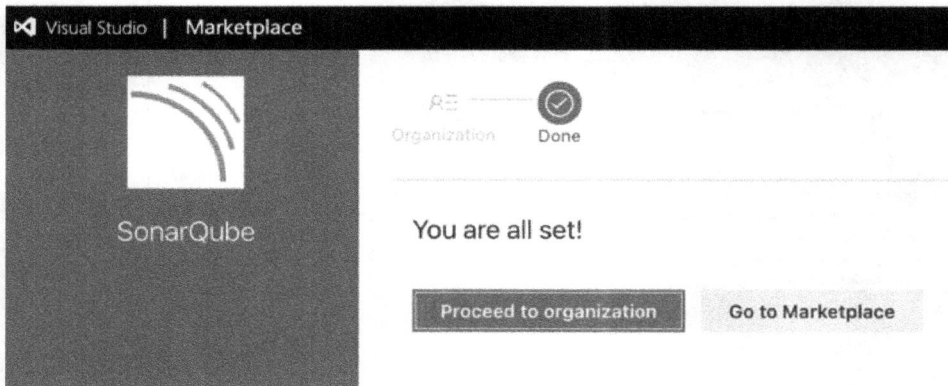

Figure 6.4 – The SonarQube extension has been installed

You will find the code for this chapter at https://github.com/PacktPublishing/Implementing-CI-CD-Using-Azure-Pipelines/tree/main/ch06.

Now that we have all the technical requirements in place, let's see how we can add some automated test runs to your build pipeline.

Including automated tests for your build

All modern applications require some sort of validation to ensure they are working correctly, regardless of the number of developers working on the code at the same time. This is where **automated tests**, which are executed right after the application is built, can validate that there is no loss of quality or bugs are introduced with the changes made.

There are many types of tests, such as **unit tests**, **integration tests**, and **load tests**, that can be executed against an application. There are also many automated testing frameworks available, depending on the programming language used to build the application and the preferences of the teams working on it.

> **Why is automated testing important?**
> Automated tests allow you to reduce the chance of releasing bugs in your applications by detecting them in the early stages of the development cycle, all while reducing the amount of time dedicated by testing teams to perform the verifications and avoiding human hours in repetitive tasks that could be used to develop more features and capabilities in your applications.

In this section, you will learn how to integrate the execution of unit tests into your build pipeline using the **NUnit test framework** and a sample C#.NET application created in .NET Core 6.0.

We will assume you're using a Visual Studio solution where you have a `CalculusService` Class Library project. The test projects are included. The following is the example code of this class in the Class Library project:

```
namespace CalculusService
{
    public class Additions
    {
        public int Add(int number1, int number2)
        {
            return number1 + number2;
        }
    }
}
```

The following corresponding unit test is defined in a separate test project. Make sure you have a reference to the `Nunit`, `NUnit3TestAdapter`, and `NUnit.Analyzers` NuGet packages:

```
namespace CalculusService.Tests
{
    [TestFixture]
    public class AdditionsTests
    {
        private Additions additions;
```

```
        [SetUp]
        public void Setup()
        {
            additions = new Additions();
        }

        [TestCase(1, 2, 3)]
        [TestCase(2, 4, 6)]
        [TestCase(5, -10, -5)]
        public void TestAdd(int number1, int number2, int result)
        {
            Assert.That(result, Is.EqualTo(additions.Add(number1,
number2)));
        }
    }
}
```

You will need to use a YAML pipeline, as shown in the following code snippet, to build and execute the automated tests:

```
trigger:
- main

pool:
  vmImage: 'windows-latest'

variables:
  solution: '**/*.sln'
  buildPlatform: 'Any CPU'
  buildConfiguration: 'Debug'

steps:
- task: NuGetToolInstaller@1
- task: NuGetCommand@2
  inputs:
    restoreSolution: '$(solution)'
- task: VSBuild@1
  inputs:
    solution: '$(solution)'
    msbuildArgs: '/p:PackageAsSingleFile=true
/p:PackageLocation="$(build.artifactStagingDirectory)"'
    platform: '$(buildPlatform)'
    configuration: '$(buildConfiguration)'
```

```
- task: VSTest@2
  inputs:
    testSelector: 'testAssemblies'
    testAssemblyVer2: |
      **\*.Tests.dll
      !**\*TestAdapter.dll
      !**\obj\**
    codeCoverageEnabled: true
    platform: '$(buildPlatform)'
    configuration: '$(buildConfiguration)'
```

The most important section of this pipeline is the last step, which uses the VSTest@2 task. This is a generic and out-of-the-box task available in Azure Pipelines for running unit and functional tests that support several test frameworks that take advantage of Visual Studio's Test Explorer. It's also important to set the codeCoverageEnabled property to true so that you can collect data that indicates how much of the code in the application is being tested.

> **Pro tip**
>
> When configuring the VSTest task and using the testAssemblyVer2 attribute, ensure that you provide a list of patterns to specifically find the test assemblies to execute tests in. Otherwise, you will find yourself with cryptic errors.

With unit tests executed and code coverage enabled, you will see the results in the **Summary** window:

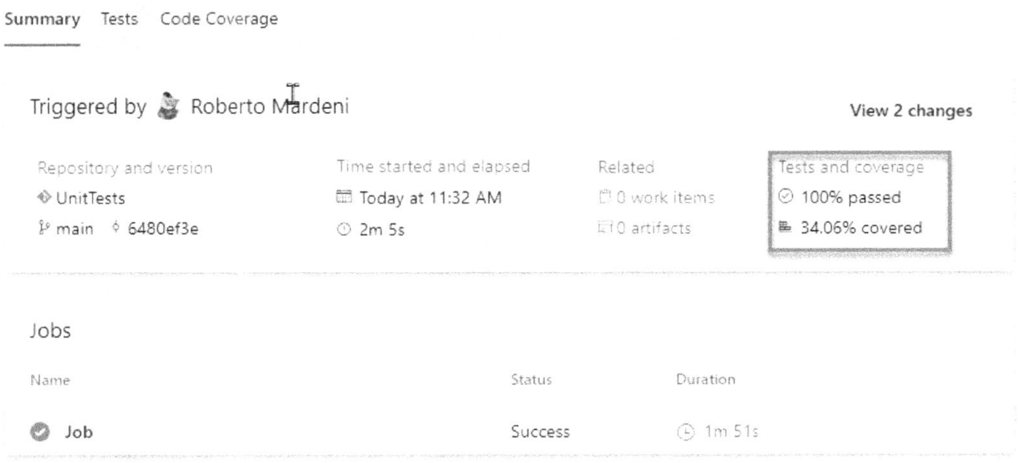

Figure 6.5 – Test and coverage results in the Summary window

The benefit of using this task is that it provides support for automatically publishing test results and UI reporting built into Azure Pipelines, as shown in the following screenshot:

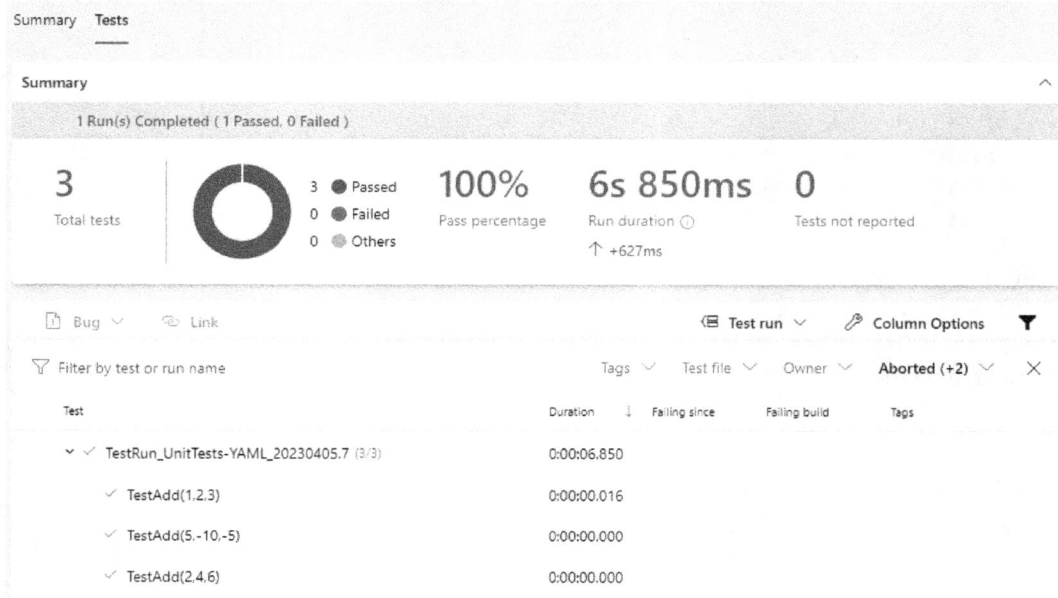

Figure 6.6 – The test results included in the Azure Pipelines run

The `VSTest@2` task also supports more advanced scenarios such as executing tests in parallel across multiple agents, which is useful when you have many tests to run, or executing UI tests, which require additional configuration in the agent that will be executing the test.

If your application is built using other programming languages, you must use the corresponding test runner and ensure the results are published in any of the formats supported by the `PublishTestResults@2` task to be able to import them and include them in the UI.

Now that we've learned about running tests in an automated way, let's learn how to increase the quality of your code.

Increasing code quality

Typically, developers are too busy to focus on the quality of their code and end up taking advantage of different automated tools to make sure they are producing the best and most secure application possible.

There are two important areas in this space to understand:

- **Static application security testing**: This allows you to detect vulnerabilities in your code
- **Software composition analysis**: This allows you to detect vulnerabilities in references to external packages and libraries used in your code

> **Why use tools to improve code quality?**
>
> Developers and testers can only do so much with the time they have available to meet timelines and work on application features. Introducing these tools early allows them to detect bugs and vulnerabilities that could otherwise be costly when the application is released to end users.

There are many well-known third-party tools you can use to scan and assess your code quality. In this chapter, we will use **SonarQube** as it is one of the most popular and easy to use. It allows developers to ensure they are producing clean code by identifying bugs and security vulnerabilities, as well as detecting common anti-maintainability patterns and duplicate code, among other things.

It comes in different pricing tiers, starting with the free Community Edition, which will be used for the examples in this chapter. If you want to get more programming language support in the tool or advanced vulnerability detection, you need one of the paid tiers.

Checkmarx, Veracode, OWASP, WhiteSource, and HP Fortify are among many others available. It is beyond the scope of this book to compare these tools, but you can certainly find plenty of comparisons online.

Follow these steps to set up SonarQube analysis on your code:

1. Configure a SonarQube project.
2. Create a service connection to SonarQube in Azure DevOps.
3. Create an Azure pipeline to analyze your code.

We'll walk through these steps in the following sections.

Configuring a SonarQube project

In your SonarQube instance, proceed to create a project from the wizard by selecting the **From Azure DevOps** option, as shown in the following figure:

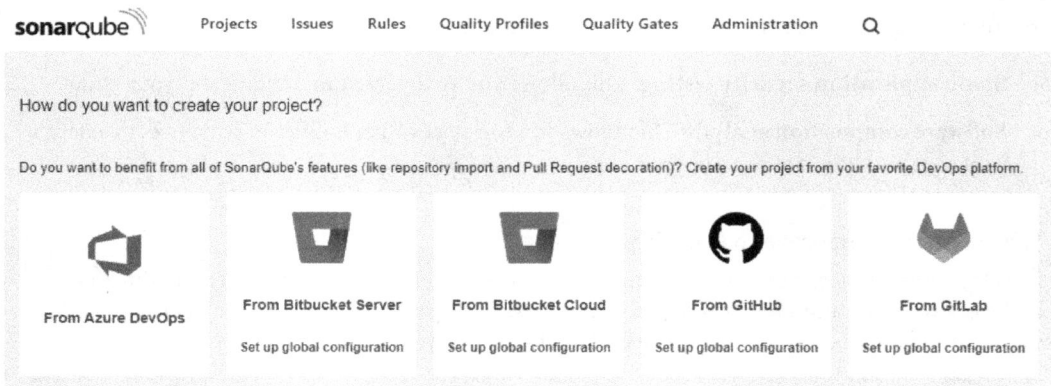

Figure 6.7 – Creating a project in SonarQube using the From Azure DevOps option

If this is the first time you are setting up a connection to Azure DevOps, SonarQube will prompt you for **Configuration name**, **Azure DevOps URL**, and **Personal Access Token** details so that it can configure the project:

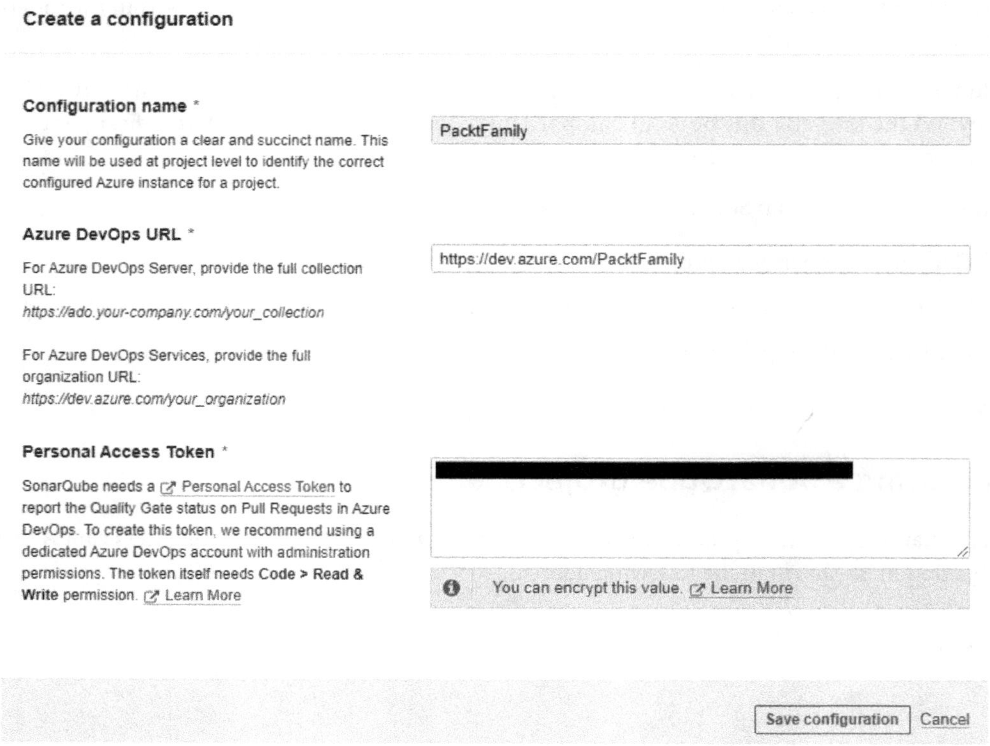

Figure 6.8 – Create a configuration

Then, select the Azure DevOps project to configure in the SonarQube side by selecting it from the available list and clicking on the **Set up selected repository** button:

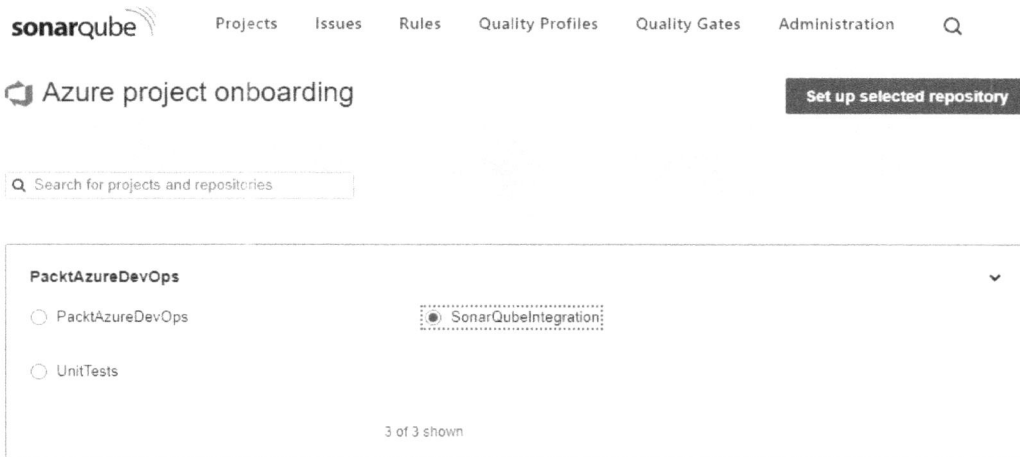

Figure 6.9 – Selecting an Azure DevOps project in SonarQube

Once you've done this, you are ready to proceed with the configuration in the Azure DevOps and pipeline side.

Creating a service connection to SonarQube in Azure DevOps

The next step is to create a **service connection** to the SonarQube instance. This will allow Azure Pipelines to use the SonarQube extension and communicate with the SonarQube instance.

You can create and manage service connections by choosing the **Azure DevOps Project Settings** option and clicking the **New service connection** button. This will show a listing where you can search for the right option by typing just a few characters in the search field. In this case, `Sonar` should list the SonarQube option:

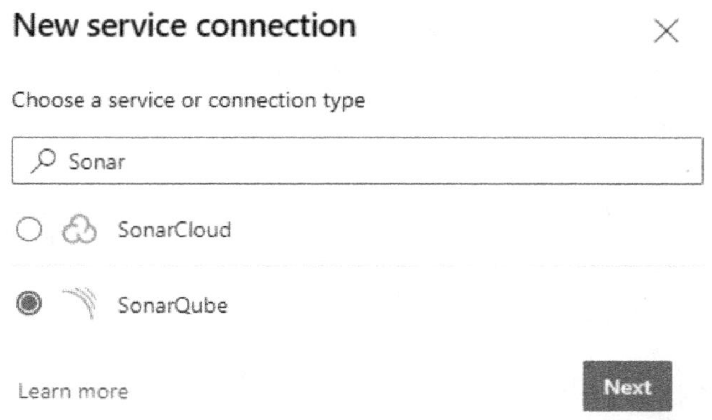

New service connection   ✕

Choose a service or connection type

🔍 Sonar

○ ☁️ SonarCloud

◉ 〰️ SonarQube

Learn more   Next

Figure 6.10 – The SonarQube service connection option

The next step is to provide details in the **Server Url**, **Token**, **Service connection name**, and **Description (optional)** boxes. Don't forget to check the **Grant access permission to all pipelines** option, unless you want to manage access to the service connection separately for each pipeline:

New SonarQube service connection   ✕

Server Url

https://83a6-2600-1700-1cf0-520f-f486-b51e-d171-bbea.ngrok.io

Uri for the SonarQube Server to connect to.

Authentication

Token

•••

Authentication Token generated through SonarQube (go to My Account > Security >
Generate Tokens)

Details

Service connection name

SonarQube

Description (optional)

Security

☑ Grant access permission to all pipelines

Learn more   Back   Save
Troubleshoot

Figure 6.11 – Service connection details for SonarQube

> **Important note**
>
> Service connections are an easy and centralized way to define the credentials needed to communicate with a service outside of Azure DevOps, removing the need to enter credentials everywhere.

Creating an Azure Pipeline to analyze your code

The next step is to include two tasks available that are in the SonarQube extension in your build pipeline. Let's look at the following pipeline:

```
pool:
  vmImage: 'windows-2019'
variables:
  solution: '**/*.sln'
  buildPlatform: 'Any CPU'
  buildConfiguration: 'Release'
steps:
- task: SonarQubePrepare@5
  inputs:
    SonarQube: 'SonarQube'
    scannerMode: 'MSBuild'
    projectKey: 'PacktAzureDevOps_SonarQubeIntegration_########'
- task: NuGetToolInstaller@1
- task: NuGetCommand@2
  inputs:
    restoreSolution: '$(solution)'
- task: VSBuild@1
  inputs:
    solution: '$(solution)'
    msbuildArgs: '/p:DeployOnBuild=true /p:WebPublishMethod=Package
/p:PackageAsSingleFile=true /p:SkipInvalidConfigurations=true
/p:PackageLocation="$(build.artifactStagingDirectory)"'
    platform: '$(buildPlatform)'
    configuration: '$(buildConfiguration)'
- task: SonarQubeAnalyze@5
```

The `SonarQubePrepare` task is needed to provide the context necessary to perform the analysis, for which the service connection, scanner mode, and project key in SonarQube must be provided. This task must be placed before any compilation tasks are executed.

The `SonarQubeAnalyze` task is responsible for performing the security scan and must be placed after all tasks that compile code. It will use the information collected since the execution of the `SonarQubePrepare` task to perform all necessary data collection and analysis for the security scan. This task will fail the pipeline if it does not pass the conditions defined in the SonarQube project.

Reviewing SonarQube analysis results

The results of the security scan will be available in the SonarQube portal. Depending on the nature of your project, you will get different quality indicators and recommendations to improve your code, as shown in the following figure:

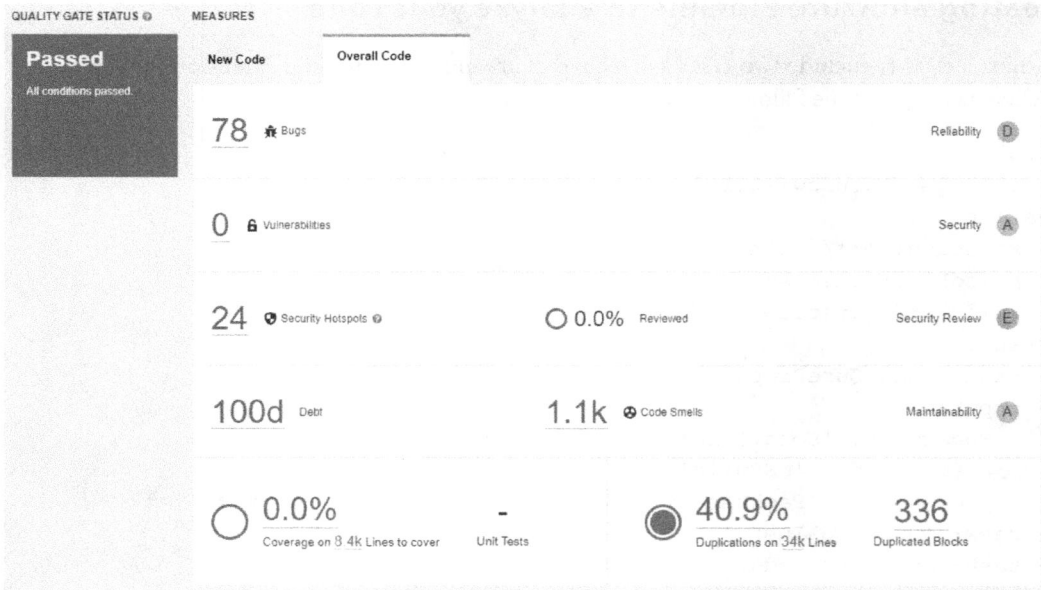

Figure 6.12 – SonarQube analysis results

> **Important note**
>
> It is possible to use the `SonarQubePublish` task to include a brief summary and link to the full report in your Azure Pipeline summary. However, this is only available for paid tiers of SonarQube.

SonarQube will provide you with insights regarding bugs, vulnerabilities, security hotspots, duplication of code, and many other issues that might be present in code, helping the developers solve these by finding them and providing instructions on how to best fix them.

Integrating this type of tool into your pipelines provides a fast feedback loop for developers to fix and mitigate risks in the applications early in the development process and avoid costly mistakes if those were deployed to a production environment.

Azure Pipelines can also be used to orchestrate the deployment of **artifacts** created in other systems, such as the popular CI/CD tool **Jenkins**. We'll look at this in detail in the next section.

Integrating with Jenkins for artifacts and release pipelines

In this section, we will walk through a simple setup demonstrating how to connect Azure Pipelines and Jenkins so that you can download an artifact generated in Jenkins and deploy it via release pipelines.

A **Jenkins Job** is like an Azure Pipeline, an automated set of steps that executes actions and can produce artifacts or perform deployments. Let's learn how to create a simple Jenkins job.

Creating a Jenkins job that produces an artifact

This scenario assumes that we have a project called `PacktFamily` in a Jenkins server, as shown in the following figure:

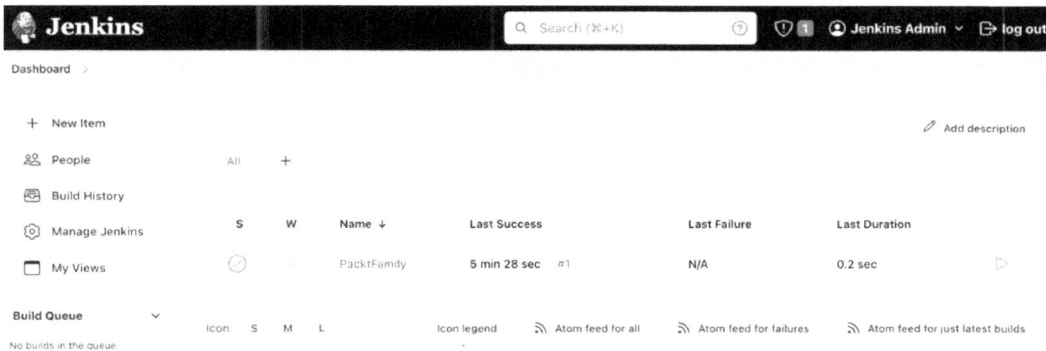

Figure 6.13 – A Jenkins instance with a PackFamily project

The configuration for the Jenkins job is very simple in this scenario, solely to demonstrate the ability the download an artifact on the Azure Pipelines side. The following figure shows the build steps for producing `artifact.txt`:

Figure 6.14 – The build step in a Jenkins job for creating the artifact

The following figure shows the post-build actions for `artifact.txt`:

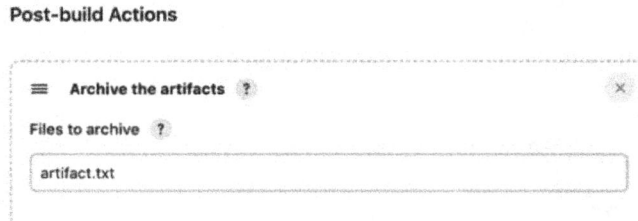

Post-build Actions

≡   **Archive the artifacts** ?   ✕

Files to archive ?

artifact.txt

Figure 6.15 – Post-build action publishing the Jenkins artifact

The execution of the Jenkins job will yield a single artifact that can be downloaded by Azure Pipelines:

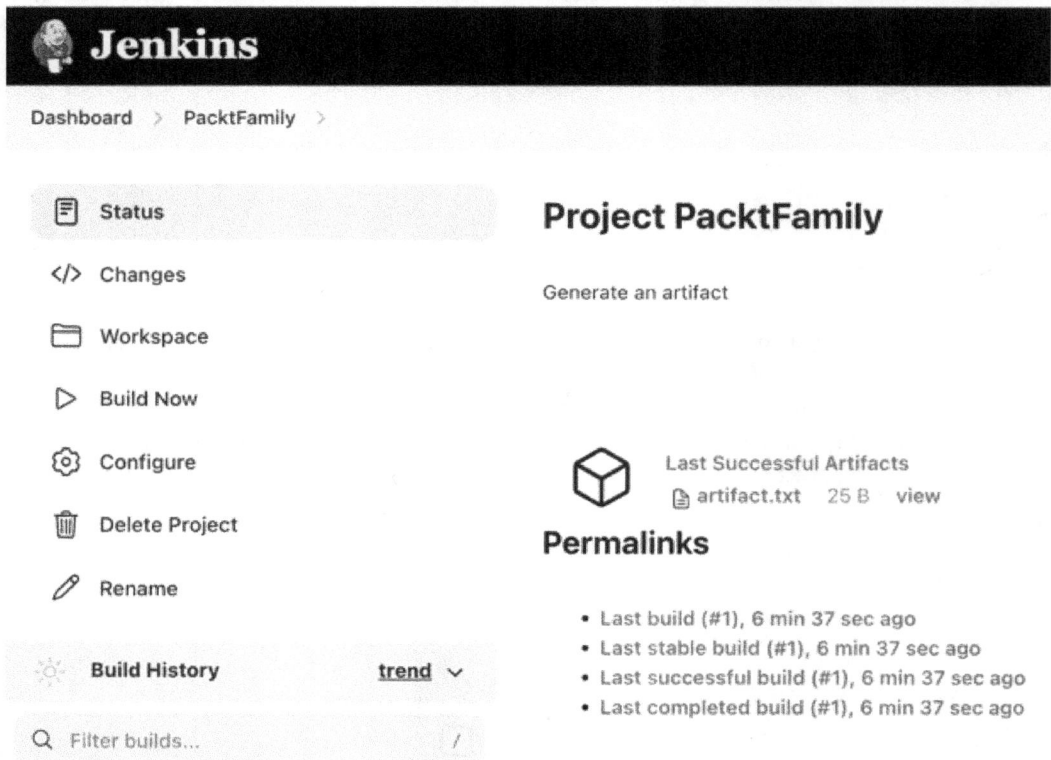

Jenkins

Dashboard > PacktFamily >

▤ Status

</> Changes

🗁 Workspace

▷ Build Now

⚙ Configure

🗑 Delete Project

✎ Rename

Project PacktFamily

Generate an artifact

Last Successful Artifacts
📄 artifact.txt   25 B   view

Permalinks

- Last build (#1), 6 min 37 sec ago
- Last stable build (#1), 6 min 37 sec ago
- Last successful build (#1), 6 min 37 sec ago
- Last completed build (#1), 6 min 37 sec ago

☀ **Build History**   **trend** ⌄

Q Filter builds...   /

Figure 6.16 – Jenkins job results and artifacts

Now that we have a Jenkins job, let's learn how to integrate Azure Pipelines with it.

Creating a service connection to Jenkins in Azure DevOps

This process is similar to what we discussed in the *Creating a service connection to SonarQube in Azure DevOps* section. In Azure DevOps' project settings, click on the **Service Connections** section and then the **New service connection** button. Use the search box to type Jenkins and click **Next**:

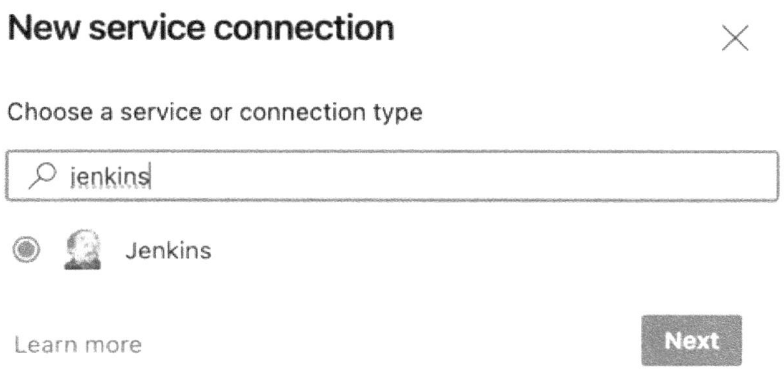

New service connection   ✕

Choose a service or connection type

🔍 jenkins

◉ 🖼 Jenkins

Learn more   **Next**

Figure 6.17 – The New service connection dialogue

Provide the **Server URL**, **Username**, **Password**, and **Service connection name** details. Don't forget to check the **Grant access permission to all pipelines** box if needed. Finally, click the **Verify and save** button to proceed:

New Jenkins service connection   ✕

Server URL

https://c231-66-176-57-252.ngrok.io/

☐ **Accept untrusted SSL certificates (optional)**
Allows the Jenkins clients to accept self-signed SSL server certificates without
installing them into the TFS service role and/or Build Agent computers.
Authentication

Username

jenkinsadmin

Username for connecting to the endpoint

Password

················

Password for connecting to the endpoint

Verify ✓ Verification Succeeded

Details

Service connection name

Jenkins

Description (optional)

Security

☑ **Grant access permission to all pipelines**

Learn more   **Back**   **Verify and save**   ⌄

Figure 6.18 – The New Jenkins service connection dialog

Now, we can proceed to create a pipeline that will use the artifact.

Creating a release pipeline to use Jenkins artifacts

Now, it is time to configure a release pipeline. You can follow these steps:

1. Navigate to **Project | Pipelines | Releases** and click **New release pipeline**. You will have the option to select a template, as shown in the following screenshot. We will start with an **empty job**:

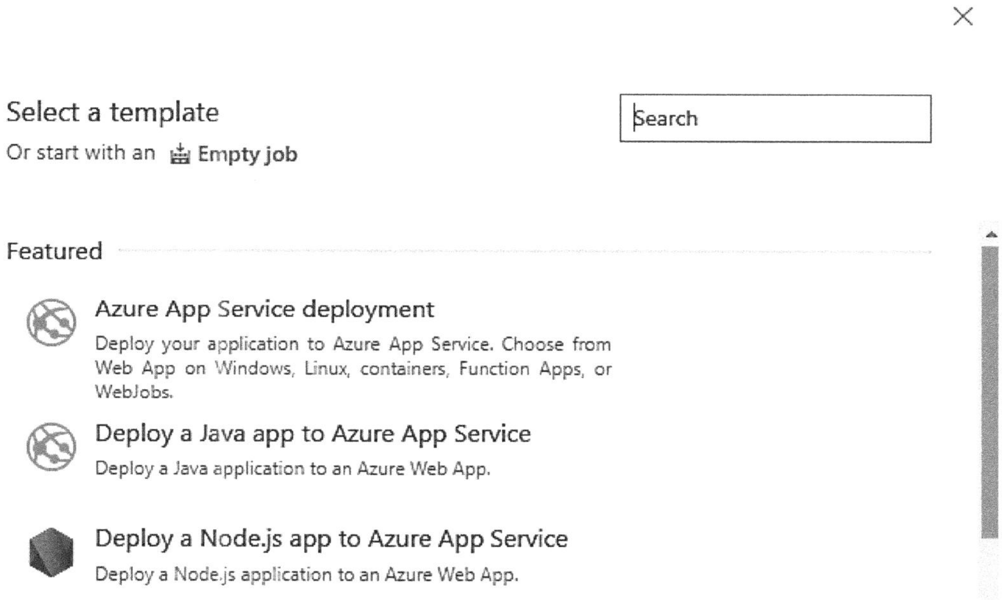

×

Select a template | Search

Or start with an 🖧 Empty job

Featured

🌀 **Azure App Service deployment**
    Deploy your application to Azure App Service. Choose from Web App on Windows, Linux, containers, Function Apps, or WebJobs.

🌀 **Deploy a Java app to Azure App Service**
    Deploy a Java application to an Azure Web App.

🧊 **Deploy a Node.js app to Azure App Service**
    Deploy a Node.js application to an Azure Web App.

Figure 6.19 – Selecting a template for a release pipeline

2. By clicking on the **Add an artifact** widget and then the **Jenkins** option, you will be able to use the previously created service connection to pick the project in Jenkins from where you will use artifacts. Just pick the service connection that matches the name you created in the previous step and then the corresponding **Source (job)** option:

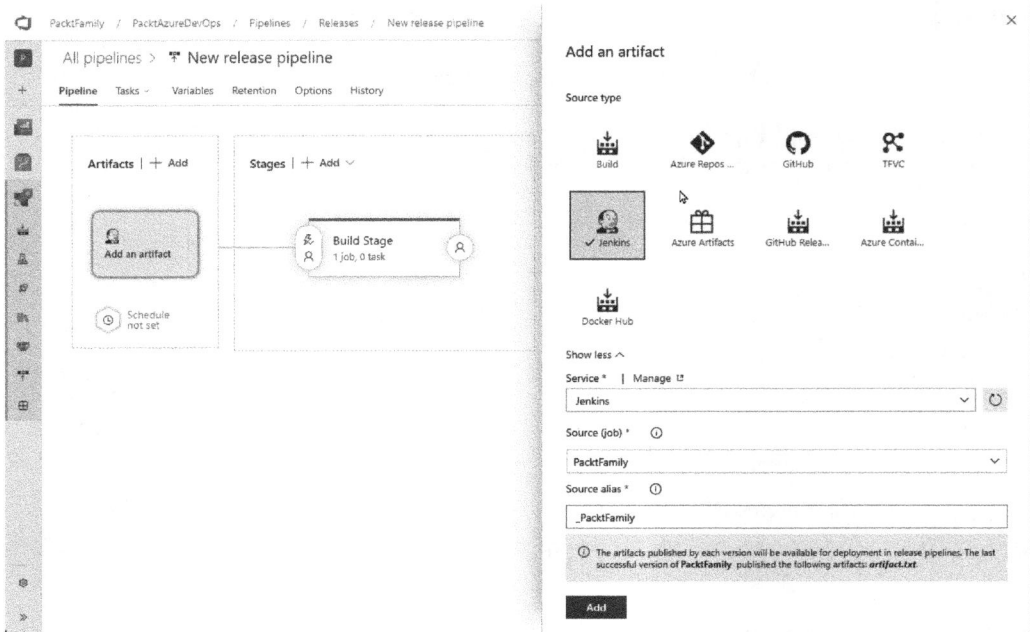

Figure 6.20 – Adding a Jenkins artifact to a release pipeline

3. You have the option to change the **Source alias** detail, which will be used as the directory where you can download artifacts once the pipeline executes. This is important when you have multiple artifacts from potentially different sources, to avoid any files from being overwritten when the pipeline executes. In this case, the default value will work.

4. Once we've done this, we can add steps to the **Deploy** stage to verify and even print out the content of the artifact. Clicking on the **1 job, 0 task** option in the **Deploy** stage will allow us to customize the pipeline. For this scenario, we will use a Linux agent. Clicking on the **Agent job** option, as shown in the following screenshot, gives us access to the **Agent selection** section. Now, we can select **Azure Pipelines** from the **Agent pool** dropdown and **ubuntu latest** from the **Agent Specification** dropdown within the **Agent selection** section:

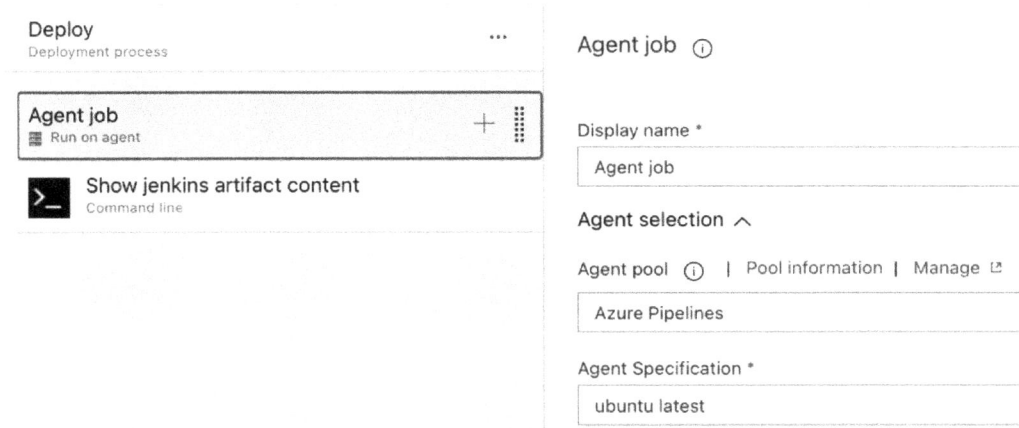

Figure 6.21 – Selecting an agent in the Deploy stage

5. Once you've done this, click on the + button on the right-hand side of the **Agent Job** section to look up the **Command line** task to add it. This task can execute a custom script in the agent and will switch to the appropriate underlying process, depending on the operating system:

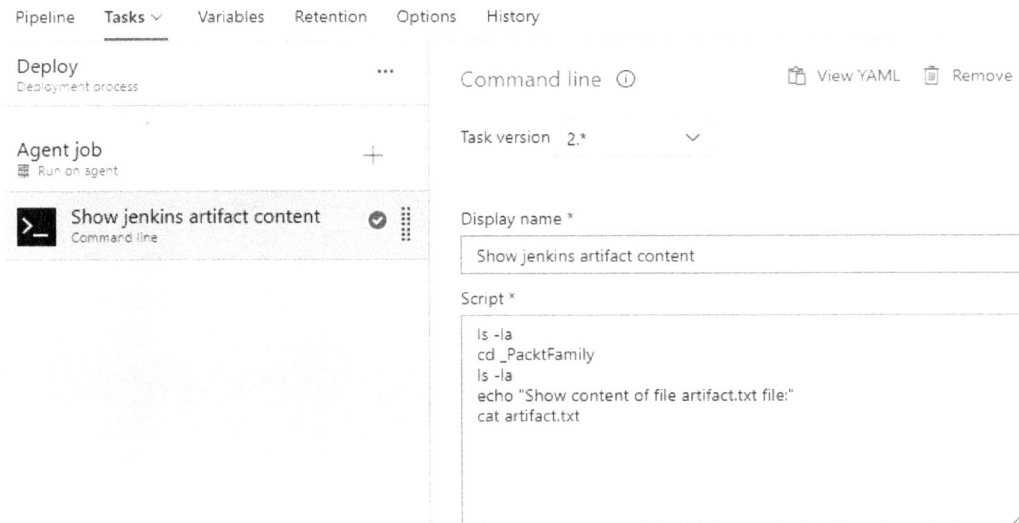

Figure 6.22 – The command-line task to list the contents of the artifact

Let's look at the script we used to show the contents:

```
ls -la
cd _PacktFamily
ls -la
echo "Show content of file artifact.txt file:"
cat artifact.txt
```

This script will do the following:

- List the contents of the current directory, which should be where the agent is running the current pipeline

- Move into the directory where the Jenkins artifacts were downloaded

- List the contents of the current directory, which should be where the Jenkins artifacts were downloaded

- Print a label indicating the contents of the file that will be displayed

- Print the contents of the `artifact.txt` file

6. Once you've saved the pipeline and created a release to execute it, you should be able to see that it effectively downloads the artifacts from Jenkins and lists the contents of the file, as shown in the following figure:

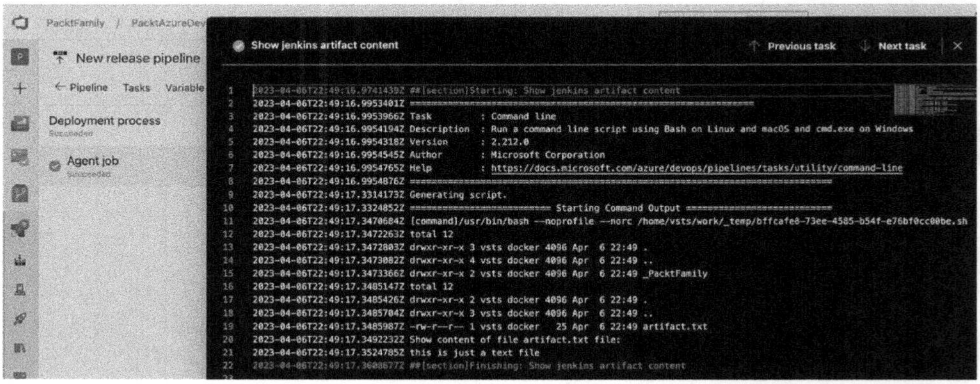

Figure 6.23 – The logs of the pipeline downloading Jenkins artifacts

> **Important note**
>
> The **Azure Pipelines** option in the **Agent selection** section provides access to Microsoft-hosted agents. These are managed by the Azure DevOps platform without you needing to manage the underlying infrastructure. Several operating systems are supported and different versions also include different tools already installed in them to facilitate building and deploying applications. You also have the option to purchase parallel jobs capacity so that you can run multiple jobs at the same time. If needed, you can also install any required software in these agents during the pipeline's execution. Just remember that these will increase the execution time.

With that, we've completed this chapter.

Summary

In this chapter, we learned about the extensibility model of Azure DevOps and how the marketplace of extensions makes it extremely easy to find additional features to include in your build and release pipelines with ease. This will speed up your ability to create build and release pipelines and integrate them with other tools. We also learned how to increase the quality of our applications by integrating automated tests and security scans to alert developers in case something breaks or introduces a vulnerability, which will reduce the amount of time needed to find bugs, fix them, and reduce security risks before you deploy your applications to the final production environments. Then, we learned how to integrate Azure Pipelines to download artifacts from another CI/CD tool and use it for deployment, which can be useful in hybrid setups where not all teams are using the same CI/CD tools. Finally, we learned about the flexibility of the Microsoft-hosted agents that are available in Azure Pipelines. This allows you to implement your CI/CD needs without having to manage the infrastructure.

In the next chapter, we will learn about monitoring Azure DevOps Pipelines, an important task that will ensure everything is working correctly and that if things go wrong, we get the visibility needed to fix them promptly.

7

Monitoring Azure Pipelines

So far, we have learned about most of the building blocks that are required to use Azure Pipelines for CI/CD needs. By the end of this chapter, you will have the skills to understand the operational aspects of running build and release pipelines efficiently, using built-in capabilities to measure the healthiness of agents, ensure timely execution of jobs, and validate that applications are running without issues after deployment.

In this chapter, we will cover the following topics:

- Understanding monitoring concepts

- Monitoring pipeline tasks and their performance

- Monitoring pipeline agents

- Measuring application quality with monitoring

But first, let's cover a few of the technical requirements for this chapter.

Technical requirements

To complete this chapter, you will need to install the *Build Quality Checks by Microsoft* Marketplace extension. Similar to the previous chapter, search for the extension in the Visual Studio Marketplace and install it in your Azure DevOps organization. You can find the extension at `https://marketplace.visualstudio.com/items?itemName=mspremier.BuildQualityChecks`. You can find the code for this chapter at `https://github.com/PacktPublishing/Implementing-CI-CD-Using-Azure-Pipelines/tree/main/ch07`.

Now that we have covered the technical requirements, let's cover the monitoring concepts you should be familiar with when working with Azure Pipelines.

Understanding monitoring concepts

When using Azure Pipelines, there are different key concepts to keep in mind when thinking about monitoring:

- **Pipeline status**: This ensures that pipelines are always running and without issues while checking for failed builds, failed tests, or errors during deployments.

- **Code quality metrics**: This involves verifying metrics such as code coverage, code complexity, and code smells to identify potential performance or functionality issues before deploying applications.

- **Security vulnerabilities**: This involves assessing and measuring security vulnerabilities in the application code, dependencies, or pipeline configuration. This helps ensure that the pipeline is secure and security risks are not introduced in the application.

- **Resource utilization**: This helps ensure that the build and release pipelines are not consuming high CPU or the memory of agents or executing the pipelines for extended periods, which diminishes their ability to run other jobs.

- **Deployment health**: This involves monitoring the deployed application to ensure it is running correctly and without connectivity, availability, or functionality issues.

- **Release cycle time**: This involves monitoring the release cycle time to ensure the application deployment time is occurring on time and any delays are identified and fixed as quickly as possible. The release cycle time is the duration it takes to release a new version of the application from its initial development phase to its deployment in production.

These concepts are critical to minimize the **time to detect** (**TTD**), **time to mitigate** (**TTM**), and **time to remediate** (**TTR**) metrics, which are used in the industry to measure the ability to deliver applications promptly and fix/recover from any issues that might occur at any time.

In this chapter, we will focus on a few of these concepts, starting with pipeline tasks and their performance.

Monitoring pipeline tasks and their performance

In this section, we will cover two different approaches to monitoring tasks and performance:

- Using the pipeline's user interface
- Using dashboards

Let's dig into the user interface first.

Using the pipeline's user interface

Metrics about the **duration** of your pipeline, jobs, and tasks are available throughout the user interface to emphasize the importance of execution time, as can be seen in the following screenshot:

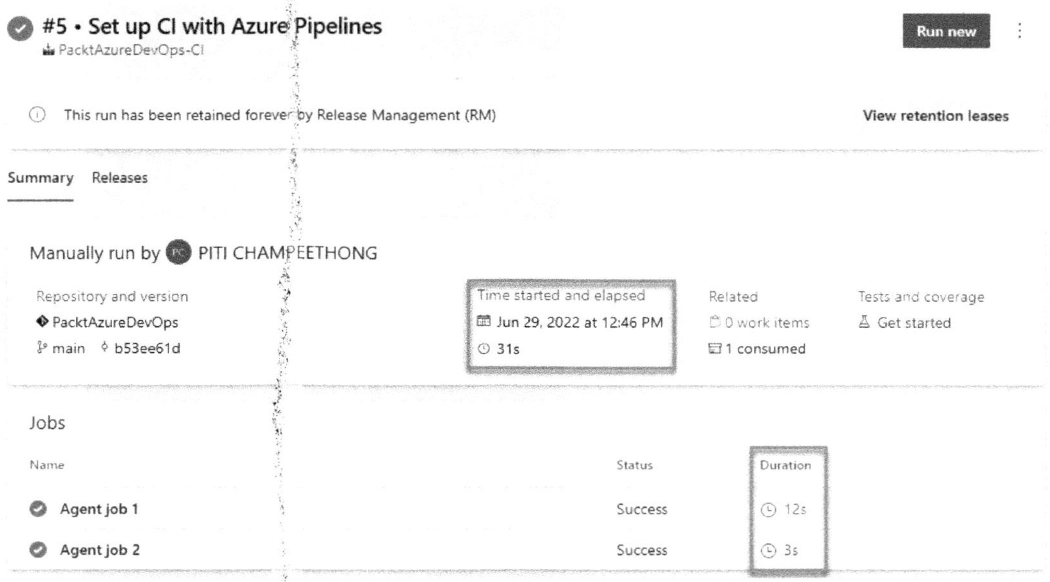

Figure 7.1 – Duration metrics in the pipeline summary

These duration metrics help you immediately understand the duration of the pipeline and all the jobs that are executed within it. You can click on each job and see the individual step duration to determine whether there are any tasks that you might need to review and improve in any way possible, as shown in the following figure:

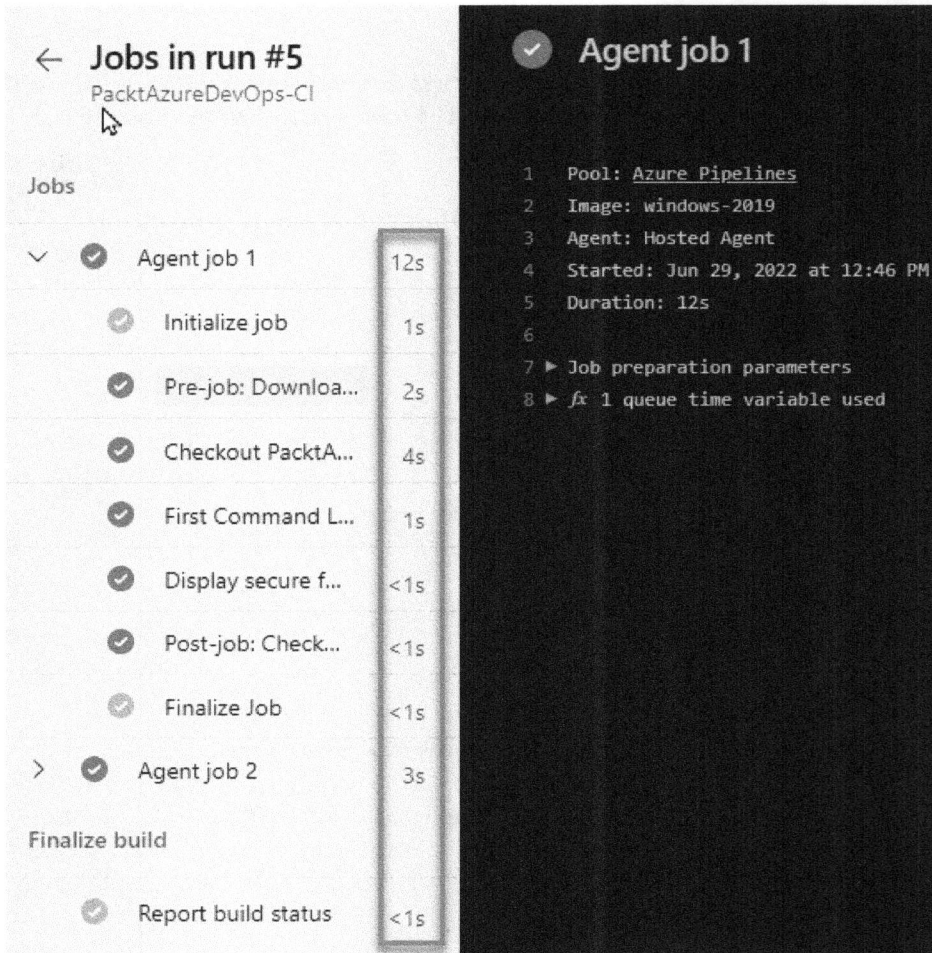

Figure 7.2 – Duration metrics in jobs

You want to always make sure the total time of execution for a build or release pipeline is the smallest possible. This ensures you are shipping software faster and deploying faster to any environment. A time increase in your pipelines can indicate that an issue has been introduced by recent changes and that you should review each task's execution time to determine whether the increase is expected and justified or whetheryou need to fix something.

You can also see the total execution time of all your pipelines by clicking the **Pipelines** option in the navigation menu and switching to the **Runs** tab, as shown in the following figure. Alternatively, you can use the filter options (highlighted in the following screenshot) to find a specific run:

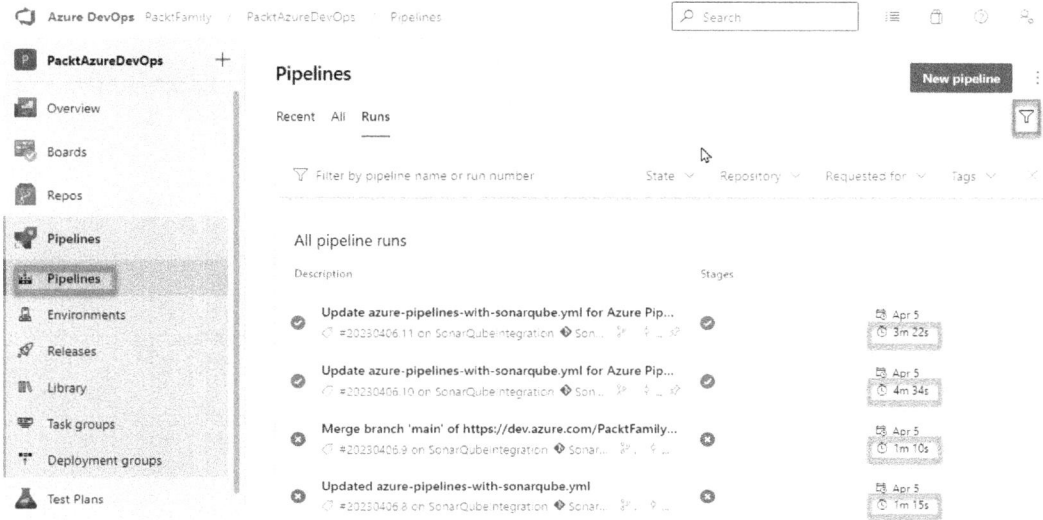

Figure 7.3 – Elapsed time of your pipelines

Looking at these metrics over time can become tedious. For this purpose, Azure Pipelines provides an **Analytics** view for each pipeline that you can access by clicking on the elements highlighted as *1*, *2*, and *3* in the following screenshots. First, navigate to **Pipelines**. From the list of recently run pipelines, select **UnitTests-YAML**:

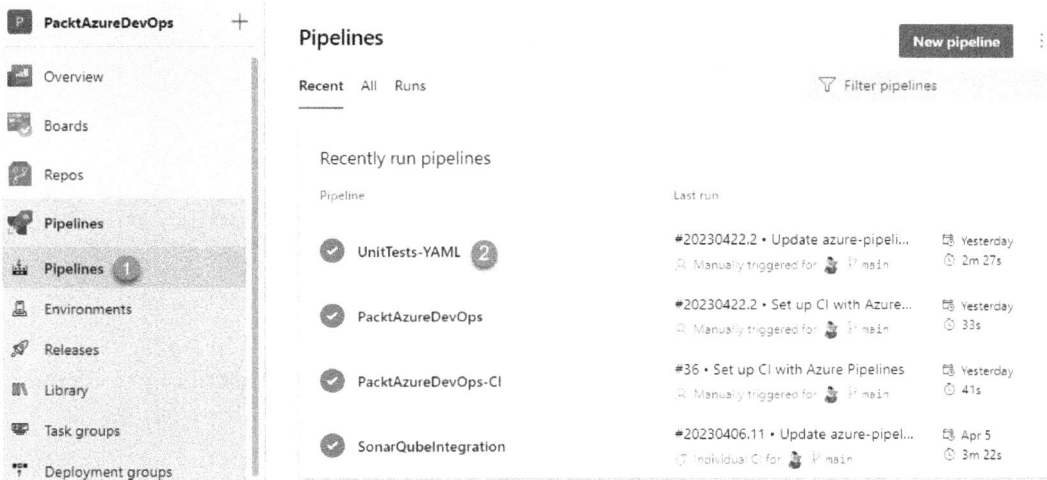

Figure 7.4 – List of all pipelines

Next, open the **Analytics** tab:

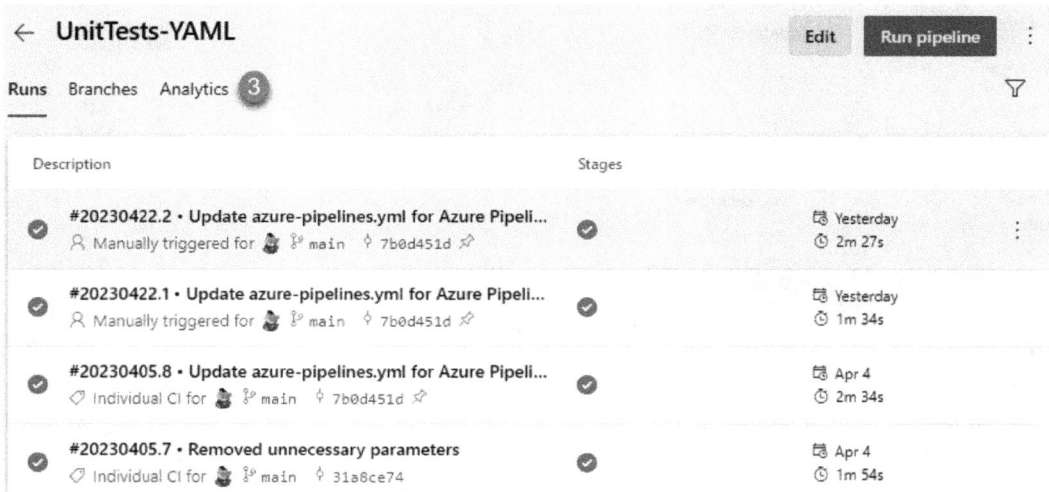

Figure 7.5 – Accessing the pipeline's Analytics view

Once the **Analytics** view loads, as shown in the following screenshot, you have three different reports that provide insights into the pipeline:

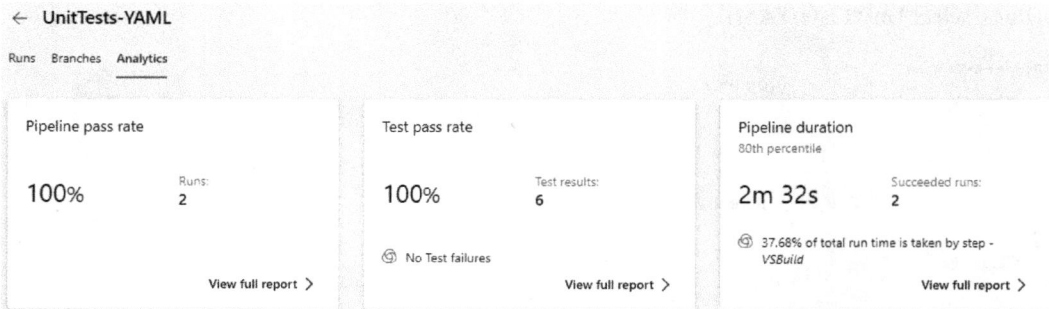

Figure 7.6 – Pipeline analytics reports

Each of the reports provides different information that is aggregated over time and can be filtered to show the last 7, 14, 30, or 180 days. As shown in the preceding screenshot, the following reports are provided:

- **Pipeline pass rate**: This reports the success or failure of the execution of the pipeline over time
- **Test pass rate**: This reports the results of unit tests over time, with the ability to show all possible test result outcomes, such as passed, failed, and inconclusive

- **Pipeline duration**: This reports the total pipeline duration over time and the top 10 steps by duration, as shown in the following screenshot:

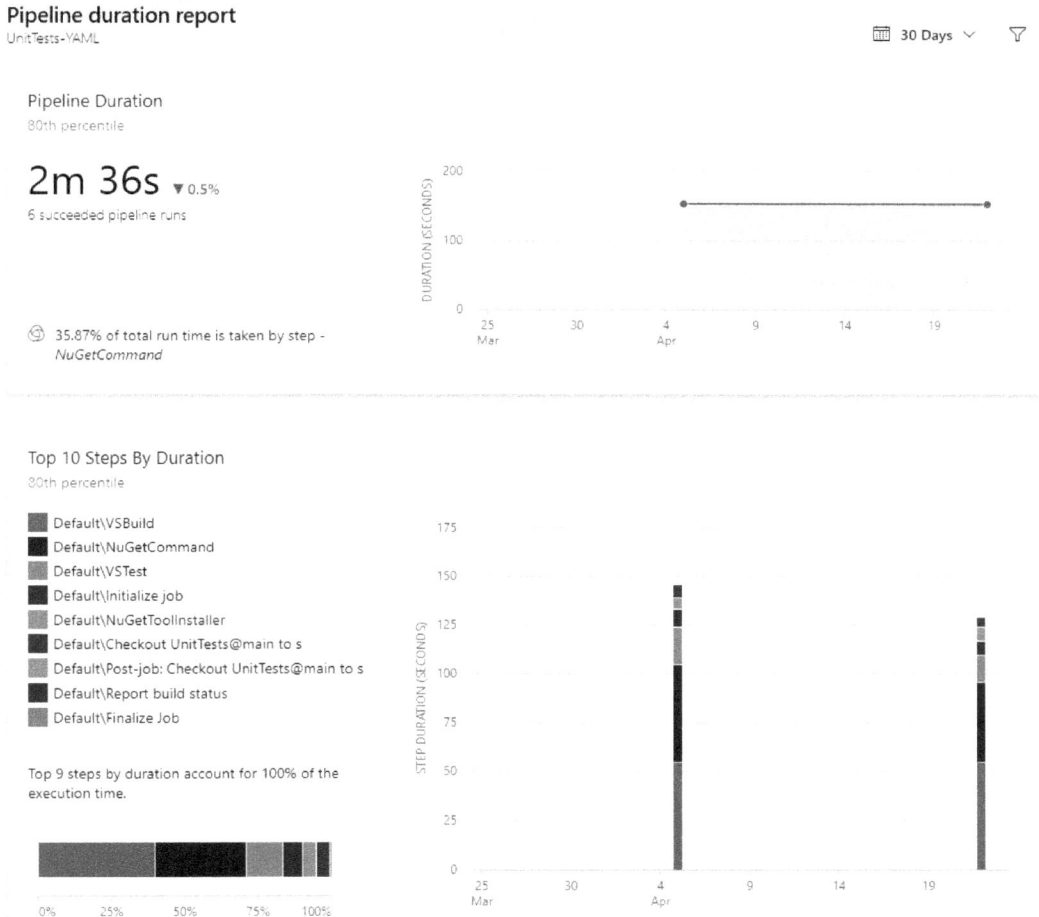

Figure 7.7 – Pipeline duration report

You can use these reports to make sure your pipelines are healthy and executing promptly. Constant and frequent revision of these reports is recommended, say weekly, to ensure that nothing unexpected has been added that has a detrimental effect on the execution time of the pipelines.

Using dashboards

Another way to monitor your pipelines is available through the **Dashboards** feature in the Azure DevOps **Overview** section of every project. You can create custom dashboards using several available widgets that display different data points that are useful at the macro level for every team member with easy and quick access.

Azure DevOps includes three out-of-the-box widgets for Azure Pipelines:

- **Build history**, which adds a tile to show a histogram of builds indicating success or failure and a link to each of them

- **Deployment status**, which adds a tile that shows a combined view of the deployment status and test pass rate across multiple environments

- **Release Pipeline Overview**, which adds a tile that allows you to view and track the status of a release pipeline

The following screenshot shows a custom dashboard named **Pipelines** that's created with all the widgets that we just discussed, showing information from different pipelines:

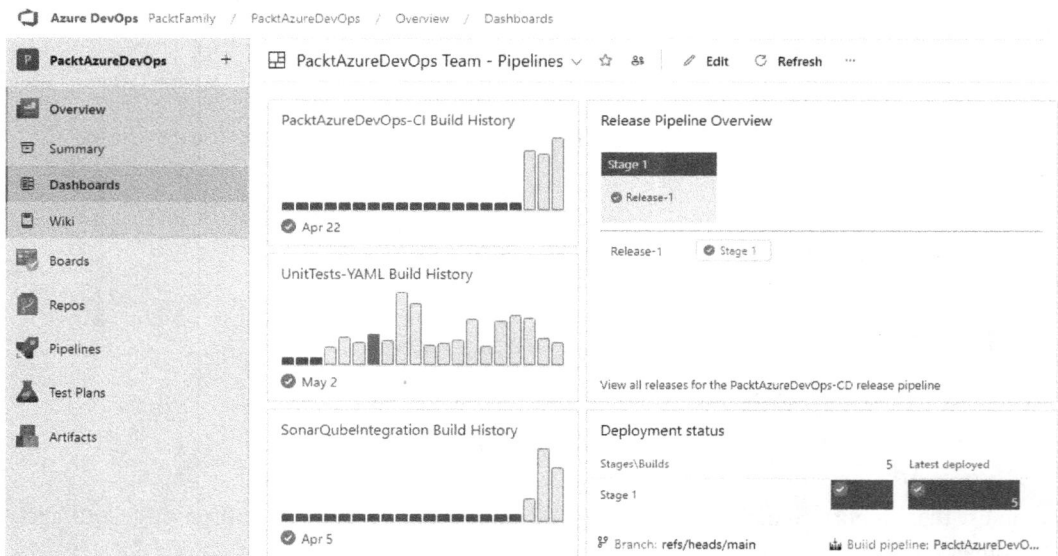

Figure 7.8 – Custom Pipelines dashboard with widgets

You can learn more about the widget catalog that's included out of the box at `https://learn.microsoft.com/en-us/azure/devops/report/dashboards/widget-catalog` and you can find more widgets by searching for them in the Visual Studio Marketplace at `https://marketplace.visualstudio.com/search?term=widgets&target=AzureDevOps`.

> **Important note**
>
> Another effective way to monitor your pipelines is via the **Azure Pipelines Microsoft Teams app** marketplace extension, which you can find in the marketplace catalog; upon clicking it, you will be taken to **Microsoft App Source** store. This application is a Teams app and is installed in your Teams tenant, which is outside the scope of this book. After you have installed it, you can configure subscriptions to the pipelines you want to get notified on pipeline status or approvals.

Now, let's learn how to monitor pipeline agents.

Monitoring pipeline agents

In Azure DevOps, pipeline agents provide some general reporting capabilities. They can be accessed by clicking **Organization settings**:

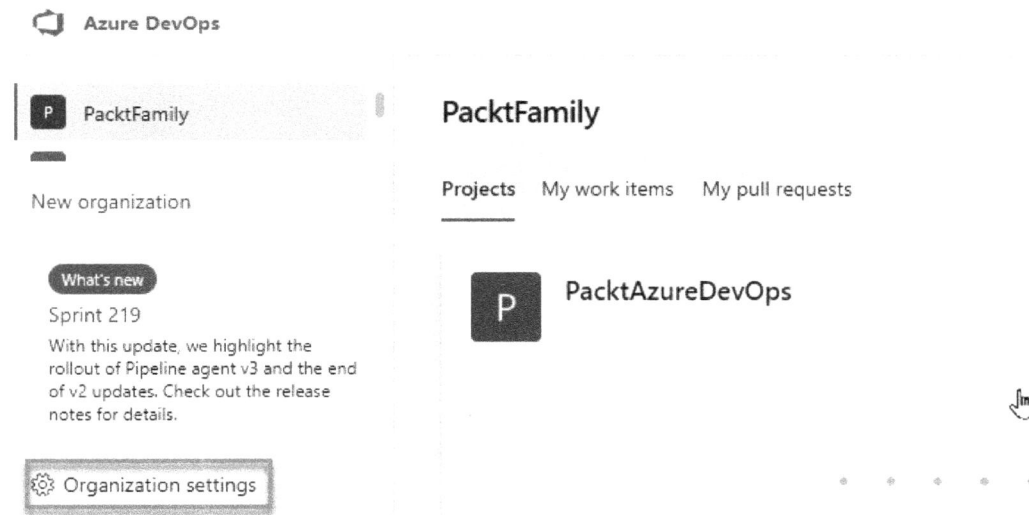

Figure 7.9 – Accessing Organization settings

Once you are inside **Organization settings**, you will have access to the **Agent pools** option in the navigation menu under the **Pipelines** section:

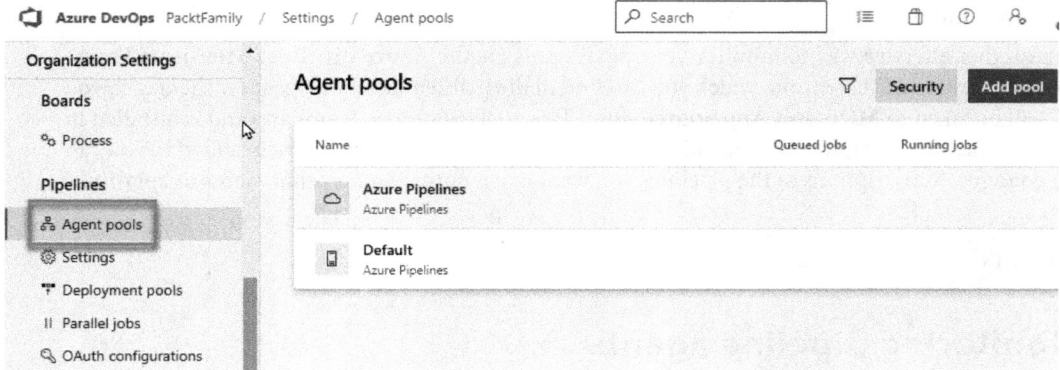

Figure 7.10 – Agent pools

Let's walk through each of the available reports.

Job runs

You can use the **job runs report** for each agent pool, which will show a summary of the jobs that are being executed, including their ID, pipeline name, project, agent specification, when they were queued, wait time, and duration:

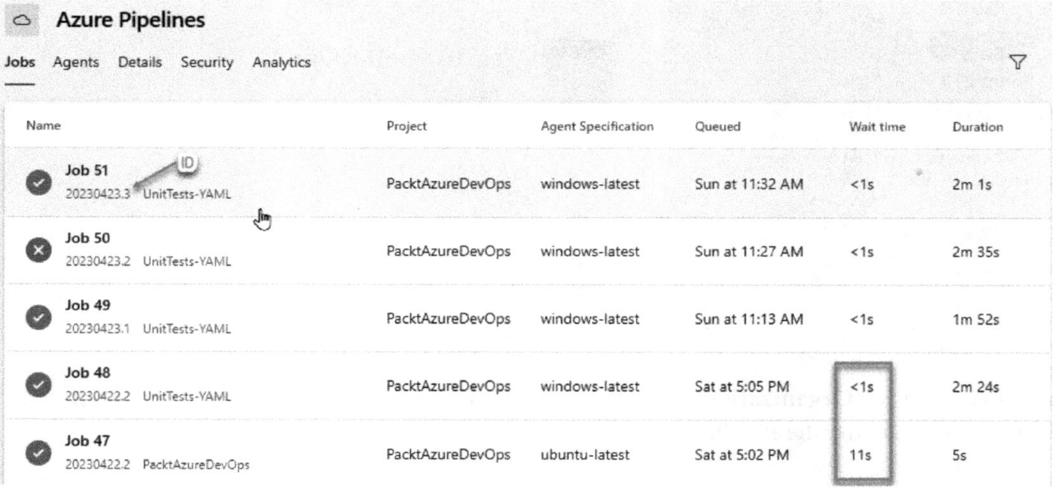

Figure 7.11 – Job reports in an agent pool

One of the most relevant pieces of data in this report is **Wait time**, which is highlighted in the preceding screenshot. This is something to pay close attention to. If this number starts increasing between jobs, it can be an indicator that you need to purchase concurrency and add additional agents.

We will cover how to purchase concurrency and the approaches to increase the number of agents later in this section. For now, let's continue to review the available reports.

Agent status

In the agent pool details, you have an **Agents** tab, which gives you details about each agent that's running, such as its name, availability, last run, current status, version, and the ability to enable/disable it.

The following screenshot shows an unavailable or **Offline** agent:

Figure 7.12 – Agent pool with an offline agent

The following screenshot shows an available or **Online** agent:

Figure 7.13 – Agent pool with an online agent

You must ensure that self-hosted agents are online and enabled when the agent pool is in use in your projects. Otherwise, pipeline jobs will be queued and never be executed if there are no agents available.

> **Pro tip**
> Always set up the agent to run as a service. This leverages the service manager of the operating system to ensure the life cycle of the agent is handled accordingly. It also improves the experience when auto-upgrading the agent.

Agent jobs

From the previous agent status report, you also have access to the **Jobs** report for a specific agent, as shown in the following screenshot:

Figure 7.14 – Agent jobs report

This is useful if you are trying to determine whether a specific agent is acting erratically or is having intermittent failures when running jobs. In this case, an agent version upgrade might be needed, dependencies tooling installed in the agent might need attention, or as a last resort, the agent should be removed along with the infrastructure where it is running and replaced with a new one.

Now, let's look at the most important report available for agent pools: analytics.

Analytics

The **Analytics** report allows us to understand aggregated usage of the agents in the pool over time with histograms indicating concurrency, queued jobs, and running jobs, as shown in the following screenshot:

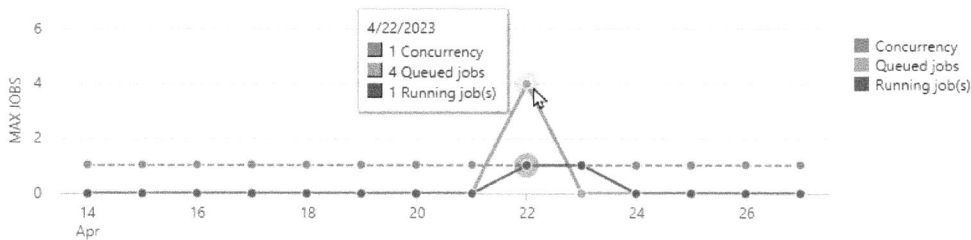

Figure 7.15 – Analytics report on an Azure Pipelines agent pool

You will notice that the report includes two histograms – one for **Public hosted concurrency** and one for **Private hosted concurrency**. The public one refers to the allowance available in Azure Pipelines for public projects, which is 10 and cannot be changed. Similarly, the private one is for private projects, for which you can purchase additional parallel jobs at an additional cost or take advantage of a self-hosted pipeline agent that's included with every Microsoft Visual Studio Enterprise subscription.

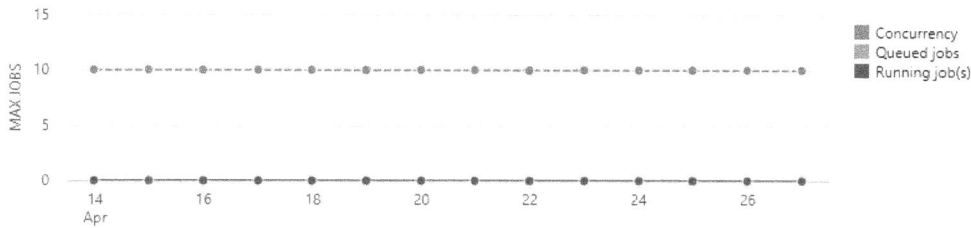

This report helps us understand when there is a need for more agents when multiple jobs are queued.

The infrequent occurrence of queued jobs could be ignored, but when they start happening often, we must consider the following aspects:

- Purchasing concurrency
- Increasing the available agents

Let's talk about purchasing concurrency first.

Purchasing concurrency

Adding concurrency to your agent pools applies to both Microsoft-hosted and self-hosted agents and the decision to increase it relies on your business need to not have wait times between job executions. This can be done by following these steps:

1. First, set up billing at the organization level, as shown in the following figure:

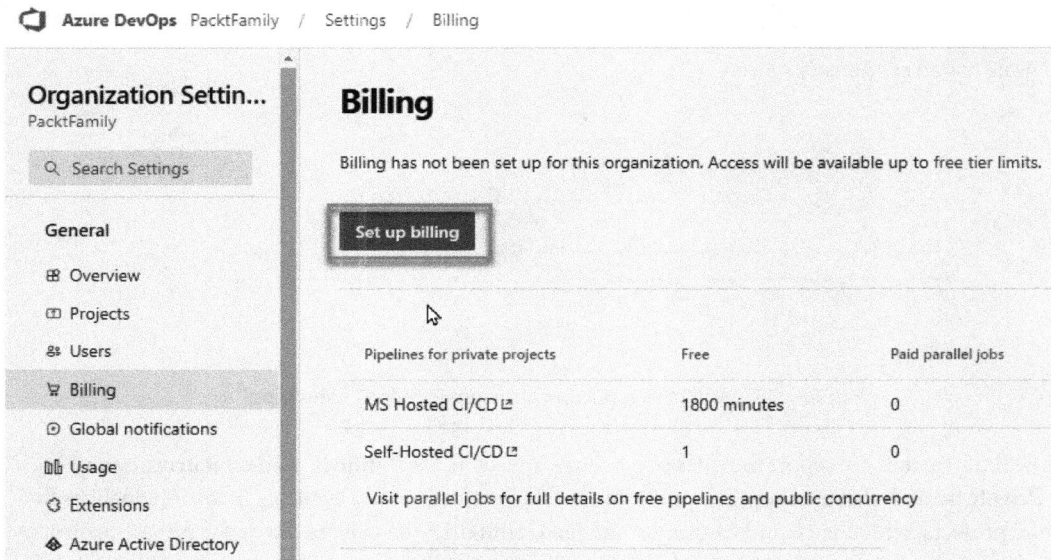

Figure 7.16 – Setting up billing for the Azure DevOps organization

2. Clicking on the **Set up billing** button will open a dialog where you can link your Azure DevOps organization with an Azure subscription, which is used to pay for services in Azure DevOps. If you have access to an Azure subscription, select it and click **Save**:

Pay-As-You-Go-17
fa8937c6-c964-4f15-8ae7-5fa9d22b6df5

⊘   Subscription is valid

+   **New Azure subscription**

Cancel   **Save**

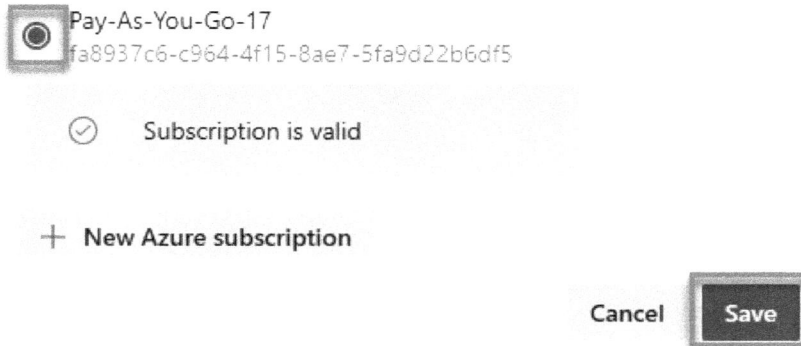

Figure 7.17 – Selecting an existing Azure subscription for billing

It is important to mention the Active Directory tenant you are logged in to is used to find Azure subscriptions you have access to and link them to the organization. You must be a member of **Project Collation Administrators Group** to complete this step.

3.   If you do not have an Azure subscription available, you will see a message similar to the following:

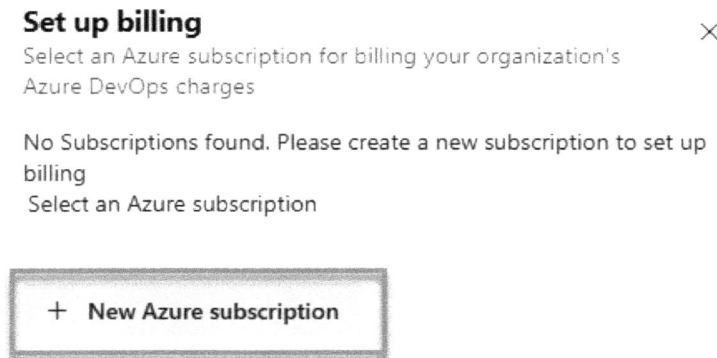

Set up billing   ✕

Select an Azure subscription for billing your organization's
Azure DevOps charges

No Subscriptions found. Please create a new subscription to set up
billing
 Select an Azure subscription

+   **New Azure subscription**

Figure 7.18 – Adding a new Azure subscription for billing

4.   You can then click on the **New Azure subscription** button to complete the steps for creating a new subscription and provide credit card details to be billed for purchases. Once billing has been configured, you will be able to purchase concurrency by going to the **Parallel jobs** option under **Project Settings**. You can increase/decrease the number of parallel jobs as needed:

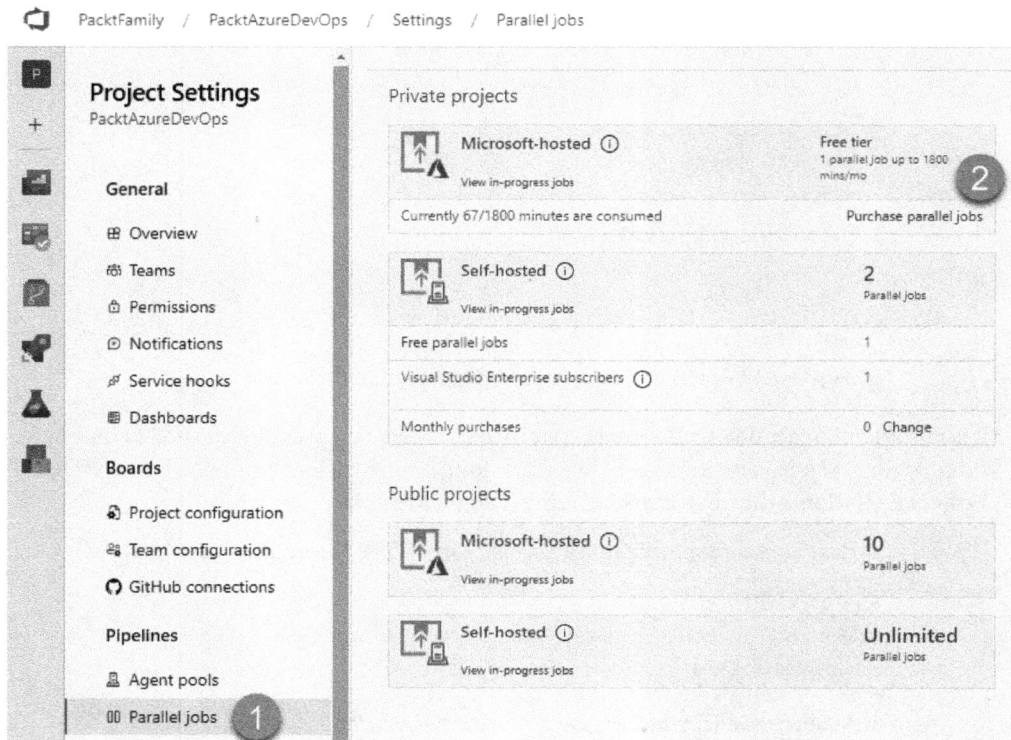

Figure 7.19 – Purchasing parallel jobs

Now that we know how to purchase concurrency, let's talk about how to increase the number of available agents.

Increasing the available agents

In the case of Microsoft-hosted agents, purchasing additional concurrency is all that is needed to immediately gain the ability to execute multiple jobs since there is no infrastructure to manage.

In the case of self-hosted agents, you have some options:

- One option was discussed in the *Setting up agent pools* section in *Chapter 1*, and it requires manually adding one agent to an agent pool.

- Another option is to use an **Azure virtual machine scale set** pool type, which is an option that becomes available when you add a new agent pool, as shown in the following screenshot. This type of agent pool automatically adds additional agents when the capacity is needed by monitoring the state of the current agents and the number of jobs in the queue every 5 minutes. You can configure the maximum number of agents and other parameters that control how each agent is handled within the pool:

Add agent pool   ✕
Agent pools are shared across an organization.

Pool type:

Select an agent pool type	⌄

Self-hosted

Azure virtual machine scale set

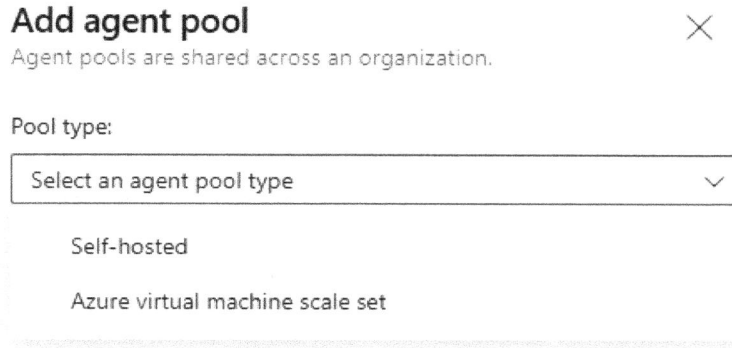

Figure 7.20 – Azure virtual machine scale set agent pool type

The **Azure virtual machine scale set** agent pool type is very useful for the following reasons:

- When you need more resources, such as CPU and memory, for the jobs to execute than the Microsoft-hosted ones and you don't want to manage the underlying infrastructure

- To simplify how the base images of these agents are maintained or the need to reimage the agents after every job execution, which this agent pool type will handle gracefully

- One last option to consider is running a **self-hosted agent** in Docker containers, which allows you to run within any container orchestrator, such as your own self-managed Kubernetes cluster or cloud-managed services such as **Azure Kubernetes Service** (**AKS**) and Amazon **Elastic Kubernetes Service** (**EKS**). In this case, you would need a custom implementation to scale the number of agents in and out based on the metrics available in the Azure DevOps REST API.

Now that we've learned how to monitor job runs, agent status, and task performance and increase concurrency and the number of agents, let's learn how to use monitoring in our pipelines.

Measuring application quality with monitoring

Azure Pipelines provides many capabilities to measure the success of your build and release pipelines. First, we'll learn how to improve the success of unit test results by measuring code quality metrics.

Code quality metrics

Putting together a comprehensive unit tests pipeline includes analyzing every data point produced by the unit test runner framework and the tasks used to execute them in the pipeline. However, typically, there are limitations in terms of the metrics that are used by the task to determine failures on something other than the actual unit tests that are executed.

For example, consider a development team that recently added unit tests to a project that has been developed for years and they are simply starting to implement tests to automate and reduce the need for manual tests.

In this scenario, the general practice is to start with a small number of automated tests and work your way up to always increase that number of tests. The only way to enforce this within your pipelines is to continually monitor the unit test results and have an automated gate to evaluate this to ensure that the number is increased on every run.

Assuming you completed the previous chapter, the following task will accomplish this scenario by adding it to the end of the YAML file for the unit test build pipeline:

```
- task: BuildQualityChecks@8
  inputs:
    checkCoverage: true
    coverageFailOption: 'build'
    coverageType: 'blocks'
    forceCoverageImprovement: true
    coverageUpperThreshold: '80'
```

You can find the complete file at `https:https://github.com/PacktPublishing/ Implementing-CI-CD-Using-Azure-Pipelines/blob/main/ch07/azure- pipelines.yml`

Let's break this down to understand what this task does:

- The `checkCoverage: true` option enables the policy that requires code coverage results to be present.

- Next, `coverageFailOption: 'build'` indicates the build will fail if code coverage is not increased when compared to the previous build.

- `coverageType: 'blocks'` indicates that analysis will be done over the number of blocks of code. Other options include `lines`, `branches`, and `custom`.

- `forceCoverageImprovement: true` will enforce that the value of the code coverage metric is always higher than that of the previous run.

- Finally, `coverageUpperTreshold: '80'` is the upper threshold for code coverage improvements. Typically, you will not strive for 100% code coverage as this implies every single line of code has a test associated with it, and in very large applications, this might not be realistic as it would require more development time. Once this value is reached, there will be no more enforcement for improvement.

Adding this to the previously configured `UnitTests-YAML` pipeline and making no additional changes will result in a failed execution because no tests are being added to increase the code coverage metric:

#20230504.1 • Update azure-pipelines.yml for Azure Pipelines
UnitTests-YAML

Rerun failed jobs   Run new   ⋮

ⓘ This run will be cleaned up after 1 month based on your project settings.

Summary   Tests   Code Coverage   Extensions

Triggered by Roberto Mardeni   View change

Repository and version	Time started and elapsed	Related	Tests and coverage
UnitTests	Today at 8:08 PM	0 work items	100% passed
main 3399084d	3m 36s	0 artifacts	34.06% covered

Errors 1

❌ The code coverage (percentage) value (36.5756%, 1647 blocks) has not improved since the last build! The previous build had a coverage val...
BuildQualityChecks

Troubleshooting failed runs

Jobs

Name	Status	Duration
❌ Job	Failed	3m 25s

Figure 7.21 – Failing a build with quality checks

Now that we've learned how to use code quality metrics to enhance our pipelines, let's see how we can improve deployments.

Deployment health

CI/CD allows you to automate every aspect of the deployment process, including validating the application in the target environment after it has been deployed. This scenario provides a mechanism to ensure that no human intervention is required to verify that a new version of an application is working as expected and no new errors or bugs have been introduced by the developers or the environment configuration.

Let's look at a simple scenario first. Here, we'll consider the **Jenkins Artifacts** release pipeline that we discussed in the previous chapter, in which we deployed an artifact from a Jenkins job.

We didn't explicitly add a step to verify that the `artifact.txt` file we expected was copied and made available to the agent. This can be addressed by adding a command-line task with a customized script, as shown in the following screenshot:

All pipelines > ⚓ Jenkins Artifacts   💾 Save   🚀 Create release   ⋯

Pipeline   **Tasks ⌄**   Variables   Retention   Options   History

Deploy Deployment process ⋯	Command line ⓘ 📋 View YAML 🗑 Remove

Agent job   +
🖥 Run on agent

>_ **Check if artifact available** ✓ ⠿
    Command line

>_ **Show jenkins artifact content**
    Command line

Task version 2.* ⌄

Display name *

Check if artifact available

Script *

```
if [[ ! -f _PacktFamily/artifact.txt ]] ; then
  echo "File _PacktFamily/artifact.txt not found"
  exit 1
fi
```

Figure 7.22 – Validating a task in a release pipeline

The script shown in the preceding screenshot works for an Ubuntu agent and verifies whether the `artifact.txt` file in the `_PacktFamily` directory exists; otherwise, it will print a message indicating that the file was not found and will exit with a return code of 1. This will be interpreted as an error since the task is always expecting a return code of 0 to succeed.

Let's look at another scenario, say the deployment of a web application or Web API, for which you can write scripts that can issue an HTTP/HTTPS request to the application, wait for the response, and validate the response code and content.

An even better scenario would be to use a UI-automated test framework and execute them as part of the release pipeline after the application has been deployed, just like the unit tests that we explored in the previous chapter. The following UI automated test frameworks are a few options to consider:

- **Open source**:

  - **Appium**: https://github.com/appium/appium

  - **Robot Framework**: https://robotframework.org/

  - **Selenium**: https://www.selenium.dev/

- **Third-party**:

  - **Cypress**: https://www.cypress.io/

  - **Sauce Labs**: https://saucelabs.com/

  - **Telerik Test Studio**: https://www.telerik.com/teststudio

In more advanced scenarios, you can use **gates** in Azure Pipelines, which give you the ability to introduce automated points of control to evaluate conditions defined based on the task used. When using release pipelines, gates are available as pre-deployment and post-deployment conditions, but when using multi-stage pipelines with environments, gates are only available as post-conditions attached to the environment.

We'll explore one of these deployment gates using Azure Monitor next.

Integration with Azure Monitor

Azure Monitor is a monitoring solution for collecting, analyzing, and responding to logs and metrics from cloud and on-premises environments. This can help you understand how your applications and services are performing and provide the ability to manually and programmatically respond to conditions that require attention to ensure said applications keep working as expected.

The integration capability in Azure Pipelines is provided via an `AzureMonitor` task, which allows you to query rules for active alerts and determine whether the deployment of a new version of an application has triggered new alerts.

In this section, you will use a readily available template for release pipelines to easily configure the Azure Monitor task.

To do this, perform the following steps:

1.   Create a new release pipeline, as shown in the following screenshot:

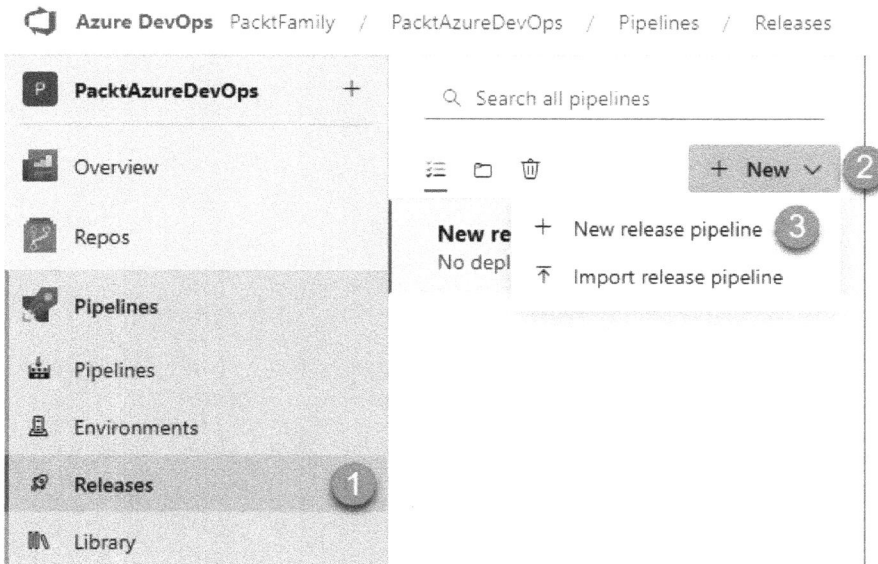

Figure 7.23 – Creating a new release pipeline

2. Find the **Azure App Service deployment with continuous monitoring** template by searching for `monitor` in the search field and clicking on the **Apply** button:

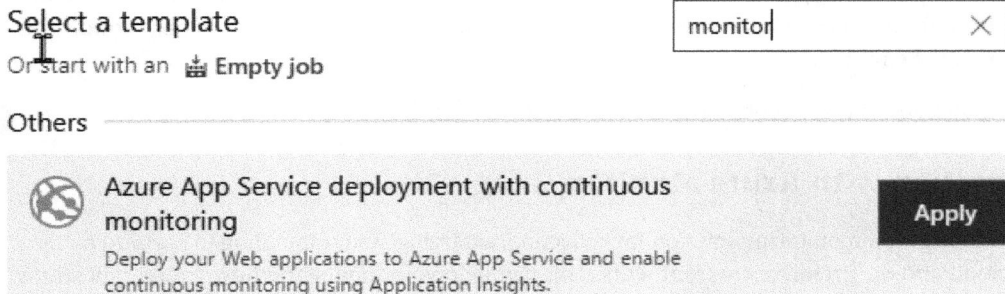

Figure 7.24 – Creating a new release pipeline from a template

3. You will end up with a stage, as shown in the following screenshot, where you must fill out the **App Service name**, **Resource Group name for Application Insights**, and **Application Insights resource name** fields:

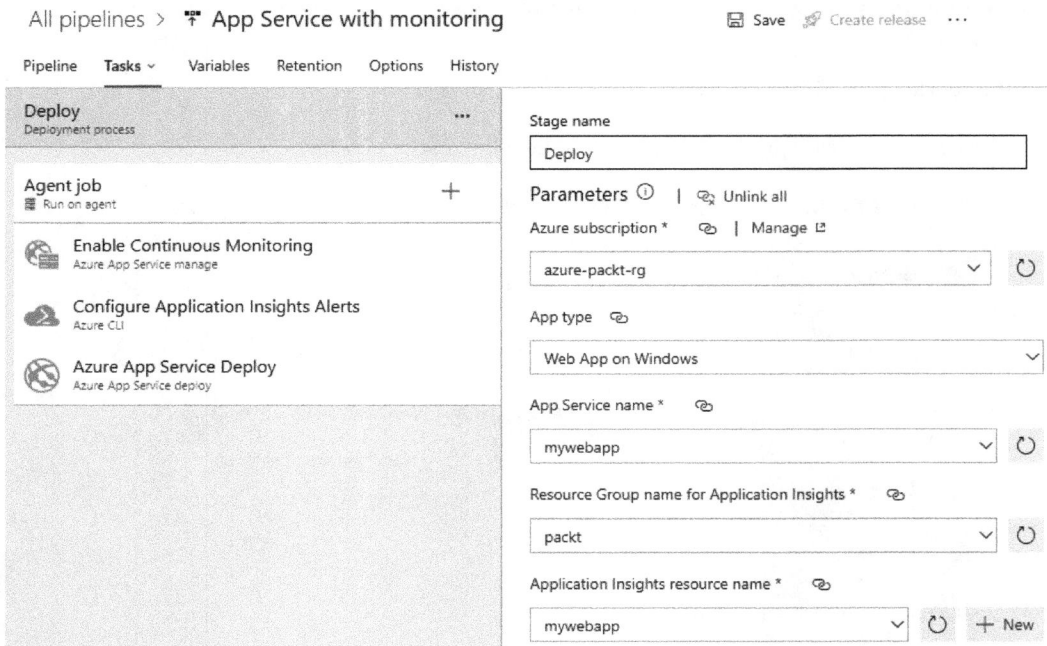

Figure 7.25 – Azure App Service deployment with continuous monitoring

4. The most important step is the **Configure Application Insights Alerts** task, which uses the Azure CLI to create four different metrics-based alerts using an inline script: `Availability_$(Release.DefinitionName)`, `FailedRequests_$(Release.DefinitionName)`, `ServerResponseTime_$(Release.DefinitionName)`, and `ServerExceptions_$(Release.DefinitionName)` with some default thresholds. You can use these defaults, adjust them, or create new alert definitions based on what is important for your application:

Inline Script *   ⓘ

```
$subscription = az account show --query
"id";$subscription.Trim("'"");$resource="/subscriptions/$subscription/resourcegroups/"+"$(Parameters.A
ppInsightsResourceGroupName)"+"/providers/microsoft.insights/components/" +
"$(Parameters.ApplicationInsightsResourceName)";az monitor metrics alert create -n
"Availability_$(Release.DefinitionName)" -g $(Parameters.AppInsightsResourceGroupName) --scopes
$resource --condition "avg availabilityResults/availabilityPercentage < 99' --description "created from
Azure DevOps";az monitor metrics alert create -n "FailedRequests_$(Release.DefinitionName)" -g
$(Parameters.AppInsightsResourceGroupName) --scopes $resource --condition 'count requests/failed >
5' --description "created from Azure DevOps";az monitor metrics alert create -n
"ServerResponseTime_$(Release.DefinitionName)" -g $(Parameters.AppInsightsResourceGroupName) --
scopes $resource --condition 'avg requests/duration > 5' --description "created from Azure DevOps";az
monitor metrics alert create -n "ServerExceptions_$(Release.DefinitionName)" -g
$(Parameters.AppInsightsResourceGroupName) --scopes $resource --condition 'count exceptions/server
> 5' --description "created from Azure DevOps";
```

Figure 7.26 – Application Insights Alerts

5. With this stage configured, you can switch to the pipeline view and click **Post-deployment conditions** to configure the gates, which in this case should already have **Query Azure Monitor alerts** enabled:

Stages | + Add ⌄

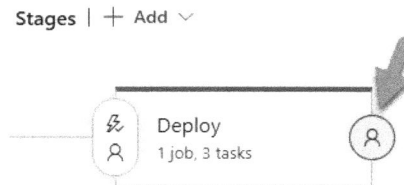

     ⚡ Deploy
      👤 1 job, 3 tasks

Figure 7.27 – Release pipeline post-deployment conditions in Stages

6. You can then adjust many of the settings related to the deployment gate accordingly, including the required **Azure subscription** and **Resource group** name and particularly **The delay before evaluation**:

Post-deployment conditions
Deploy

   ᐞ Post-deployment approvals   (⬤) Disabled
Select the users who can approve or reject deployments to this stage

   →] Gates ∧   (⬤) Enabled
   Define gates to evaluate after the deployment.
   Learn more

   The delay before evaluation ⓘ

   | 5 | | Minutes ∨ |

   Deployment gates ⓘ   + Add ∨

      🕐 Query Azure Monitor alerts   (⬤) Enabled 🗑

      Query Azure Monitor alerts ⓘ

      Task version | 1.* ∨ |

      Display name *

      | Query Azure Monitor alerts |

      Azure subscription * ⓘ | Manage ↗

      | azure-packt-rg   ∨ | ↻

      Resource group * ⓘ

      | packt   ∨ | ↻

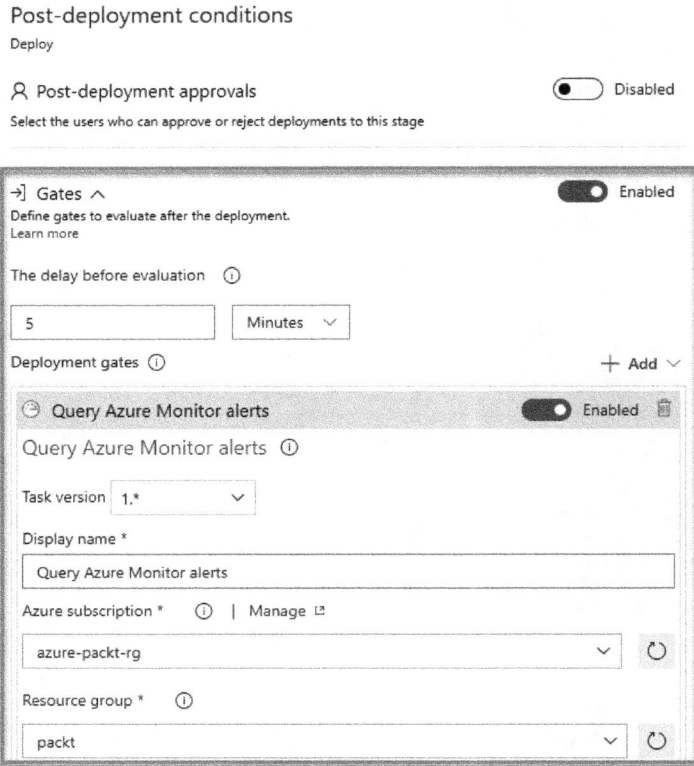

Figure 7.28 – The Query Azure Monitor alerts gate

7.   With this in place, you can rely on Azure Pipelines to execute the gate after the deployment steps are completed to verify that the monitoring alerts have been configured and provide a visual indicator:

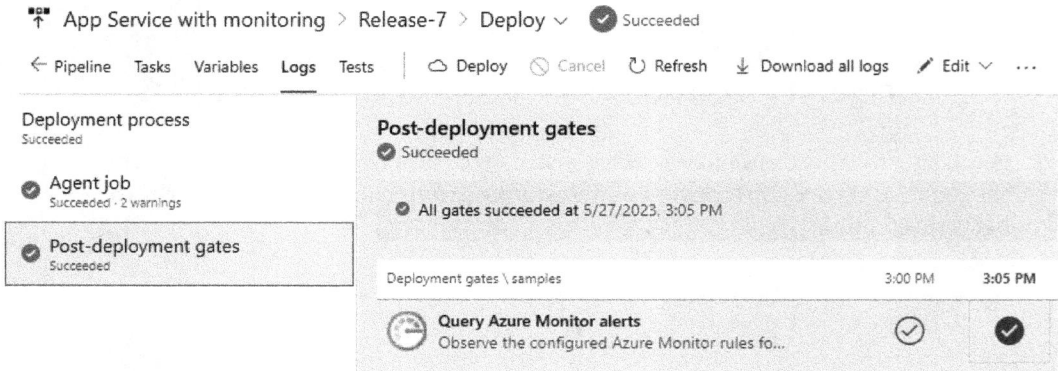

📤 App Service with monitoring > Release-7 > Deploy ∨ ✅ Succeeded

← Pipeline Tasks Variables **Logs** Tests | ☁ Deploy ⊘ Cancel ↻ Refresh ⬇ Download all logs ✎ Edit ∨ ...

Deployment process   **Post-deployment gates**
Succeeded   ✅ Succeeded

✅ Agent job
Succeeded · 2 warnings   ✅ All gates succeeded at 5/27/2023, 3:05 PM

✅ Post-deployment gates
Succeeded   Deployment gates \ samples   3:00 PM   3:05 PM

                                 🕐 **Query Azure Monitor alerts**   ⊘   ✅
                                   Observe the configured Azure Monitor rules fo...

Figure 7.29 – Successful release with deployment gates

In a complete scenario, you would combine automated tests in this release pipeline to ensure that the monitoring alerts are evaluated based on the tests being executed against the recently deployed application.

8. Finally, you could use third-party application performance monitoring tools with the corresponding available marketplace extensions:

    -   **Dynatrace**: `https://marketplace.visualstudio.com/items?itemName=AlmaToolBoxCE.DynatraceIntegration`

    -   **Datadog**: `https://marketplace.visualstudio.com/items?itemName=Datadog.datadog-monitors`

Now that you've learned about deployment gates, we've come to the end of this chapter. Let's summarize what we've learned so far.

Summary

In this chapter, you learned about monitoring concepts to consider for your CI/CD projects and different ways to monitor your pipeline tasks, their performance, and how to build dashboards with graphical widgets to understand behavior over time and even integrate with a collaboration tool to get real-time notifications. You also learned how to monitor your job runs, task performance, and agents, when to purchase concurrency, and different options on how to increase the number of agents to ensure that pipelines execute promptly. Lastly, you learned how to measure quality in your pipelines by taking advantage of code quality metrics, application runtime checks, and application monitoring tools.

In the next chapter, we are going to learn how to deploy infrastructure automatically with automation using Azure Pipelines.

8

Provisioning Infrastructure Using Infrastructure as Code

Previously we covered CI/CD topics related to building, testing, packaging, and deploying applications. In this chapter, we will learn how to provision and configure the destination where deployment will be taking place using automation, the benefits of such a process, and a few tips and tricks while we're at it. You will understand why this is important, and even required, in these times when we need to deliver fast and with quality.

In this chapter, we will dive into this by covering the following topics:

- Understanding **Infrastructure as Code (IaC)**
- Working with **Azure Resource Manager (ARM) templates**
- Working with **AWS CloudFormation**
- Working with **Terraform**

Let's take care of the technical requirements first.

Technical requirements

Depending on which section you are interested in completing in this chapter, you will need the following software installed on your workstation. You will find the code for this chapter in the GitHub repository at `https://github.com/PacktPublishing/Implementing-CI-CD-Using-Azure-Pipelines/tree/main/ch08`.

Installing Azure tools

The Azure CLI is a cross-platform command-line tool to connect to Microsoft Azure and execute commands to create, update, or destroy resources. Depending on the **operating system** (**OS**) of your workstation, you can choose the appropriate installation method at `https://learn.microsoft.com/en-us/cli/azure/install-azure-cli`. Since the instructions for each OS are different, it's up to you to complete the installation.

Once installed, run the `az version` command and you'll see a response like this:

```
PS C:\Users\user> az version
{
  "azure-cli": "2.48.1",
  "azure-cli-core": "2.48.1",
  "azure-cli-telemetry": "1.0.8",
  "extensions": {}
}
```

You can choose any text editor you wish to work with. However, we recommend **Visual Studio Code** (**VS Code** for short), as it is one of the most popular editors in the community; it's free and provides a large number of community-supported extensions for many programming languages and other tools, especially supporting **ARM templates**, **AWS CloudFormation**, and **Terraform**, as we will see in the following sections.

To install VS Code, proceed to `https://code.visualstudio.com/`. From there, you will see options to install it based on your workstation's OS.

Additionally, you must install the ARM Tools VS Code extension from `https://marketplace.visualstudio.com/items?itemName=msazurermtools.azurerm-vscode-tools`.

Installing AWS tools

The AWS CLI is a cross-platform command-line tool to connect to **Amazon Web Services** (**AWS**) and execute commands to create, update, or destroy resources. Instructions to install it based on your OS can be found at `https://docs.aws.amazon.com/cli/latest/userguide/getting-started-install.html`.

Once installed, run the `aws --version` command in a shell and you'll see a response like this:

```
PS C:\Users\user> aws --version
aws-cli/2.11.18 Python/3.11.3 Windows/10 exe/AMD64 prompt/off
```

Additionally, you must install the **AWS Toolkit** VS Code extension from `https://marketplace.visualstudio.com/items?itemName=AmazonWebServices.aws-toolkit-vscode`.

Installing Terraform tools

The Terraform CLI is a cross-platform command-line tool to execute a variety of subcommands such as `plan`, `apply`, or `destroy`, which we will cover later in this chapter. You can install it using the instructions provided at `https://developer.hashicorp.com/terraform/tutorials/azure-get-started/install-cli#install-terraform`. Once installed, run `terraform version` in a shell and you'll see a response like the following:

```
PS C:\Users\user> terraform version
Terraform v1.4.6
on windows_amd64
```

Additionally, you can install the HashiCorp Terraform VS Code extension from `https://marketplace.visualstudio.com/items?itemName=HashiCorp.terraform`.

Installing the Terraform Marketplace extension

The **Terraform Marketplace extension** for Azure DevOps must be installed. You can find it at `https://marketplace.visualstudio.com/items?itemName=ms-devlabs.custom-terraform-tasks`.

Access to an Azure account

You must have access to an Azure account to complete the steps in this chapter. If you don't have one, you can create a free one at `https://azure.microsoft.com/en-us/free/`.

Access to an AWS account

You must have access to an AWS account to complete the steps in this chapter. If you don't have one, you can create a free one at `https://aws.amazon.com/free`.

Now that we have taken care of all the technical requirements, let's walk through what it means to automate infrastructure using code.

Understanding IaC

In the past, infrastructure was typically provisioned and configured manually with manually documented steps and/or a combination of scripts. This made the whole process error-prone and slow.

In the same way that you use a rigorous process for your application code, you should practice that for your infrastructure. The purpose of this approach is to make deployments repeatable and immutable, reduce the chances of error, and accelerate the deployment process by avoiding/eliminating any human interaction whenever possible.

IaC is the practice of codifying and storing in source control a descriptive model that defines and deploys all the infrastructure needed to run your applications and any supporting dependencies. It can encompass network configuration, load balancers, virtual machines, and any other application or data services your application architecture requires to operate and is applicable to on-premises data centers and cloud provider platforms.

The best way to picture all this working together is as follows:

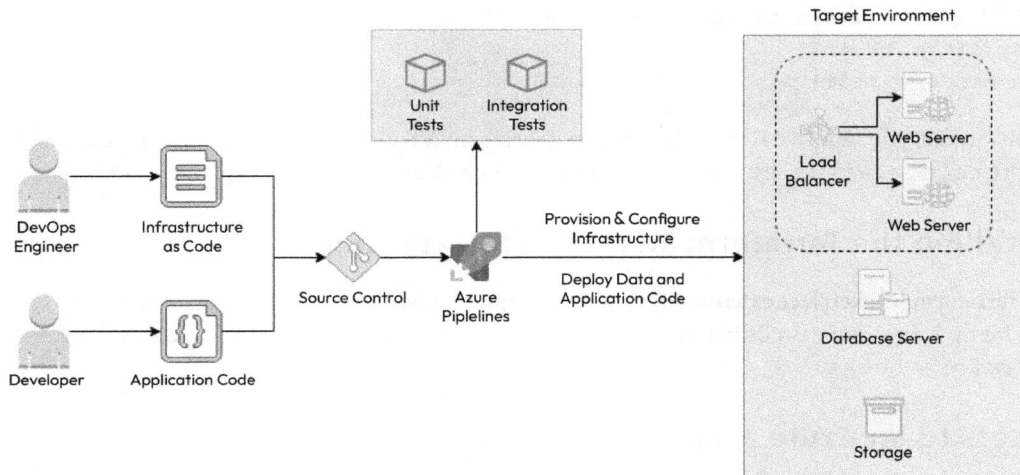

Figure 8.1 – CI/CD incorporating IaC

Let's jump into working with ARM templates now and see how we can do this on the Microsoft Azure cloud platform.

Working with ARM templates

ARM templates are one of the IaC options available to deploy infrastructure in Azure, Microsoft's cloud platform available in many regions around the world.

Microsoft also provides other tools such as the Azure CLI, Azure PowerShell, and a newer, domain-specific language called Bicep that uses a declarative syntax to deploy resources. You can also use the Azure portal, a web-based UI that provides access to all your resources in Azure and the ability to create, update, and delete resources.

ARM templates are JSON files with the following structure:

```
{
  "$schema": "https://schema.management.azure.com/schemas/2019-04-01/
deploymentTemplate.json#",
```

```
    "contentVersion": "",
    "apiProfile": "",
    "parameters": {   },
    "variables": {   },
    "functions": [   ],
    "resources": [   ],
    "outputs": {   }
}
```

ARM templates can define required and optional input **parameters**, **variables** that can be calculated for use within the template, user-defined **functions** that you can use within the template in addition to built-in ones, **resources** that define all properties to configure for one or more resources, and **outputs** that can contain properties or values calculated from the deployed resources. ARM templates can also be nested to logically separate your services.

In this section, we will focus on how to deploy ARM templates using Azure Pipelines, without getting into the details of how to create them, as that is outside the scope of this book.

Deploying ARM templates comes down to the following steps:

1. Creating a service principal in Azure
2. Creating a service connection to Azure
3. Creating an ARM template
4. Validating the ARM template
5. Deploying the ARM template

Let's start by creating a service principal in Azure.

Creating a service principal in Azure

A **service principal** is a type of identity in Azure used by applications, services, and automation tools to provide fine-grained control to access resources and perform actions based on roles.

> **Important note**
> This section assumes that you are logged in to Azure using the `az login` command.

We can create a service principal with the following Azure CLI command:

```
az ad sp create-for-rbac -n azure-pipelines --role Contributor
--scopes /subscriptions/<subscription-id>
```

You would replace `<subscription-id>`, which should be a GUID-like value from the Azure portal. Setting the scope at the subscription level is fine for testing purposes, but in your ideal setup, you will want to restrict it further – say, to a resource group:

```
/subscriptions/<subscription-id>/resourceGroups/<name>
```

Once you execute this command, you should get a response similar to the following:

Figure 8.2 – Service principal details

You must handle the service principal details as a secret. This is sensitive information that provides access to your Azure environment. You will need this for the next steps.

Now let's see how to create the service connection to Azure using the service principal.

Creating a service connection to Azure

As seen in previous chapters, integrating with external services in Azure Pipelines requires a service connection. This can be done by following these steps:

1. Navigate to **Project Settings | Pipelines | Service connections**, where you will click on the **New service connection** button:

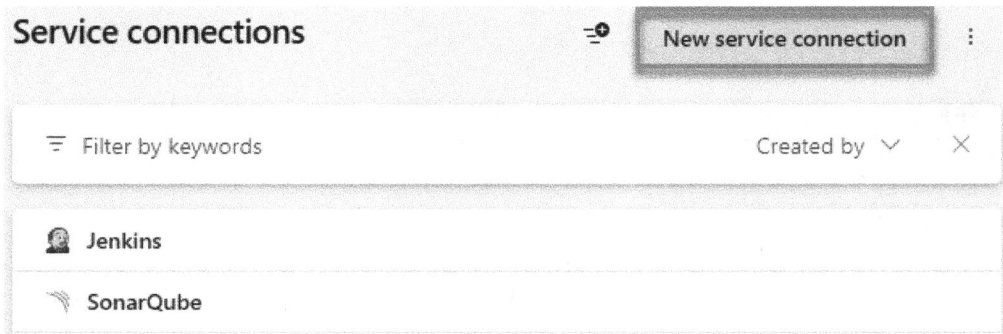

Figure 8.3 – New service connection

2. From here, you will select the **Azure Resource Manager** option and click the **Next** button:

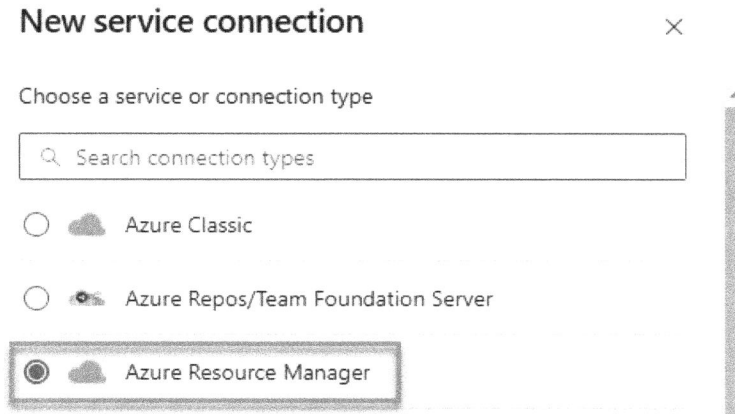

Figure 8.4 – Selecting the service connection type

3. Next, select an authentication method, using the **Service principal (manual)** option, and click the **Next** button:

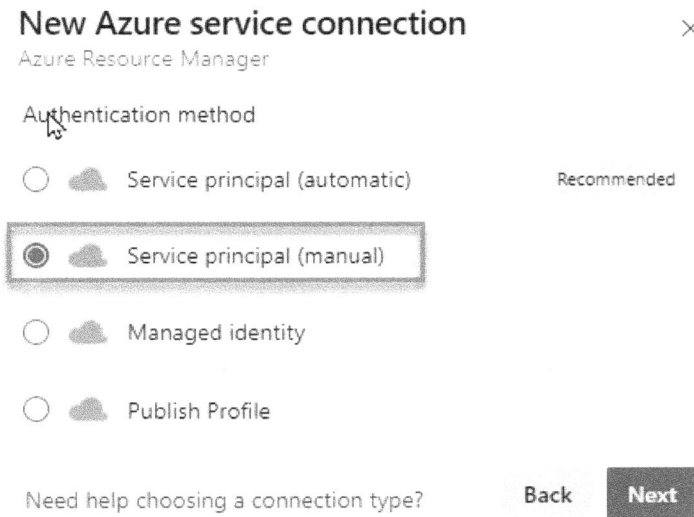

Figure 8.5 – Select an authentication method for the Azure service connection

Although the preceding screenshot shows **Service principal (automatic)** as **Recommended**, this is only from the standpoint of a user just getting started with Azure and Azure Pipelines. When you have many Azure subscriptions and resource groups, that authentication method makes it difficult to complete the setup.

4. The next step allows you to enter the details for the following parameters:

- **Subscription Id**: You can obtain this from the Azure portal

- **Subscription Name**: Also obtained from the Azure portal

- **Service Principal Id**: `appId` from the previous section, as shown in *Figure 8.2*

- **Service principal key**: `password` from the previous section, as shown in *Figure 8.2*

- **Tenant ID**: `tenant` from the previous section, as shown in *Figure 8.2*

- **Service connection name**: Use `azure-packt-rg`

Once you have entered all these values, you have the option to click on the **Verify and save** button, which will test that a connection to Azure can be established and store the service connection details.

Now that we have a service connection, let's move on to creating an ARM template.

Creating an ARM template

There are many ways to create ARM templates:

- From scratch, by following the reference documentation at `https://learn.microsoft.com/en-us/azure/azure-resource-manager/templates/quickstart-create-templates-use-visual-studio-code`

- From the Azure portal, by downloading a template before deployment from the Marketplace or an existing resource group, as described at `https://learn.microsoft.com/en-us/azure/azure-resource-manager/templates/export-template-portal`

- Modifying an existing sample template at `https://learn.microsoft.com/en-us/samples/browse/?expanded=azure&products=azure-resource-manager`

In this chapter, for simplicity, we will use a modified version of an existing sample that deploys an Azure App Service resource. You can find the template at `https://github.com/PacktPublishing/Implementing-CI-CD-Using-Azure-Pipelines/blob/main/ch08/azure/azuredeploy.json`

This ARM template will deploy two resources:

- An **Azure App Service plan**, which defines the pricing tier, OS, and other platform-level capabilities

- An **Azure App Service web app**, which defines the application-level stack, such as the PHP runtime and version

Now let's see how we can validate this template using Azure Pipelines.

Validating the ARM template

In your build or CI pipelines, you should consider validating your templates to ensure the format is correct. For this purpose, the **ARM** template deployment task is available, as shown in the following code snippet:

```
# ARM Template Validation
trigger:
- main
pool:
  vmImage: ubuntu-latest
steps:
- task: AzureResourceManagerTemplateDeployment@3
  inputs:
    deploymentScope: 'Resource Group'
    azureResourceManagerConnection: 'azure-packt-rg'
    subscriptionId: $(AzureSubscriptionId)
    action: 'Create Or Update Resource Group'
    resourceGroupName: 'packt'
    location: 'East US'
    templateLocation: 'Linked artifact'
    csmFile: 'azure/azuredeploy.json'
    deploymentMode: 'Validation'
```

> **Important note**
> YAML is a very strict language that is whitespace-sensitive and case-sensitive. When working with YAML files, make sure that you are using an editor that properly handles these requirements and make sure you are aware of any issues when formatting the content.

Let's break this down to the different parameters:

- `deploymentScope` dictates the layer to which this will be applied. Possible values are `Management Group`, `Resource Group`, and `Subscription`. These are different types of governance layers within the Azure platform and the templates follow different schemas.

- `azureResourceManagerConnection` is a reference to an existing service connection of type `Azure Resource Manager`.

- `subscriptionId` is the GUID value for the Azure subscription ID. In this case, you can see it as being a reference from a variable using the `$(AzureSubscriptionId)` notation. We will see how to create this in the next section, *Creating a pipeline variable*.

- `action` indicates whether a resource group will be created, updated, or deleted.

- **resourceGroupName** is the name of the resource group in the Azure target for the deployment. If the action is set to **create** or **update** and the resource group does not exist, it will be created by the task.

- **location** is any of the existing Azure regions available for deployment.

- **templateLocation** indicates whether the ARM template file will be available as **Linked artifact** or **URL of the file**. In the latter case, it must be a fully qualified URL.

- **csmFile** is the path to the ARM template file, required when **templateLocation = 'Linked artifact'** is set. Otherwise, you would use **csmFileLink**.

- Finally, **deploymentMode** indicates how to treat the deployment. In this case, the **Validation** value will only perform validation of the file format. We will talk about the other values this accepts in the next section.

Creating a pipeline variable

To create the variable used in this task, perform the following tasks:

1. Click on the **Variables** button on the Azure Pipelines edit screen:

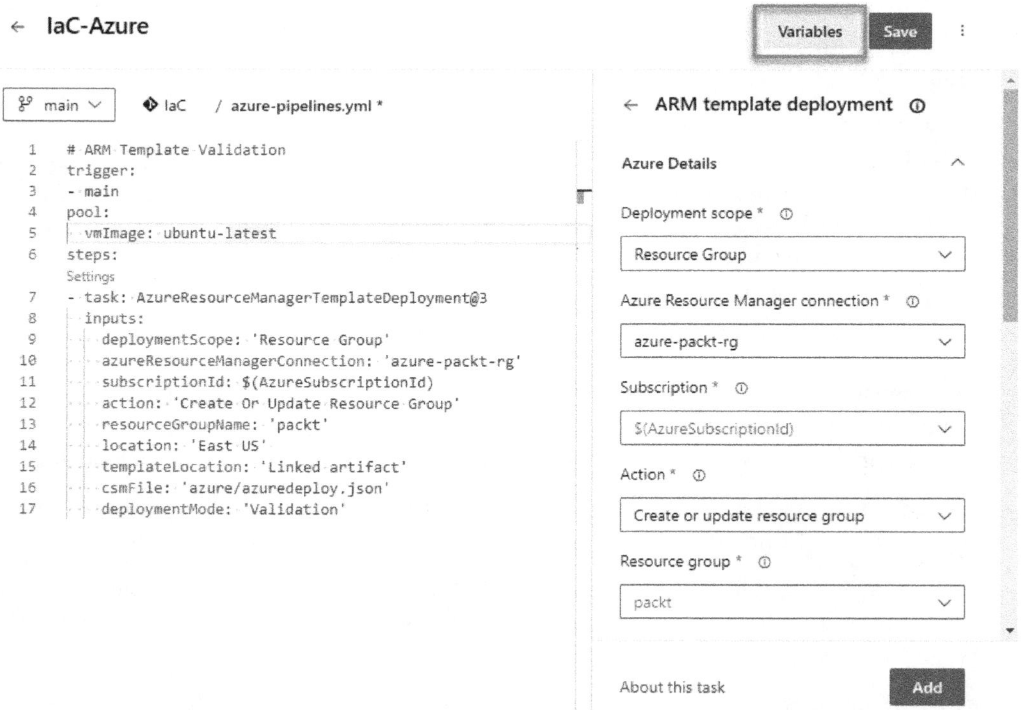

Figure 8.6 – Accessing pipeline variables

2. Click on the **New variable** button to define your first variable:

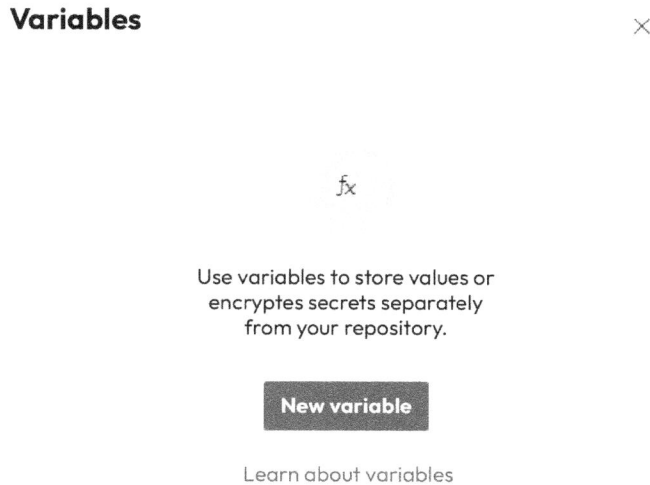

Variables   ✕

fx

Use variables to store values or
encryptes secrets separately
from your repository.

New variable

Learn about variables

Figure 8.7 – Adding a new variable

3. You then proceed to fill in **Name** and **Value**. In our case, this is `AzureSubscriptionId` and the subscription ID you obtained from the Azure portal, respectively. Also make sure to check the **Keep this value secret** option, to store it securely and make it impossible for it to be visible during pipeline execution. Clicking **OK** on this screen stores the value temporarily:

← **New variable**

Name

AzureSubscriptionId

Value

•••••••••••••••••••••••••••••

☑ Keep this value secret

☐ Let users override this value when running this pipeline

To reference a variable in YAML, prefix it with a dollar sign and enclose it in parentheses. For example: `$(AzureSubscriptionId)`

To use a variable in a script, use environment variable syntax. Replace `.` and space with `_`, capitalize the letters, and then use your platform's syntax for referencing an environment variable. Examples:

Batch script: `%AZURESUBSCRIPTIONID%`
PowerShell script: `${env:AZURESUBSCRIPTIONID}`
Bash script: `$(AZURESUBSCRIPTIONID)`

To use a secret variable in a script, you must explicitly map it as an environment variable.

Learn about variables   Cancel   OK

Figure 8.8 – New variable with secret value

4. You must click **Save** on the next screen to ensure the variable is stored in the pipeline:

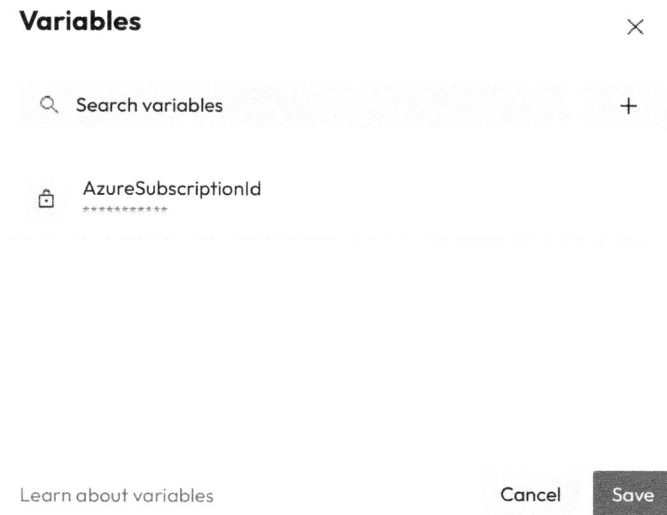

Variables   ×

Search variables   +

AzureSubscriptionId
****

Learn about variables   Cancel   Save

Figure 8.9 – Saving pipeline variables

Once you have all of this in place, the pipeline should run and validate the ARM template successfully.

> **Important note**
> When deploying ARM templates for the first time, you must ensure that the corresponding Azure resource providers have been registered in the subscription or you will get an error. For the template used in this section, you must have the `Microsoft.Web` resource provider registered. You can accomplish this by running the following Azure CLI command and waiting for it to complete:
> ```
> az provider register -namespace Microsoft.Web --wait
> ```

Now that we have learned how to validate the template, let's see how to deploy it.

Deploying ARM templates

Switching from validation to deployment requires changing the value of the `deploymentMode` property to either `Incremental` or `Complete`.

The `Incremental` deployment mode tells ARM that the resources in the template will be created if they don't exist or updated to match the template if already present. Any other resources in the resource group not defined in the template will be ignored.

Likewise, if the `Complete` deployment mode is used, ARM will ensure that the resource group only contains the resources defined in the template, create those that don't exist, update the existing ones to match, and delete any not defined in the template.

Here is what that looks like:

```
# ARM Template Deployment
trigger:
- main
pool:
  vmImage: ubuntu-latest
steps:
- task: AzureResourceManagerTemplateDeployment@3
  inputs:
    deploymentScope: 'Resource Group'
    azureResourceManagerConnection: 'azure-packt-rg'
    subscriptionId: $(AzureSubscriptionId)
    action: 'Create Or Update Resource Group'
    resourceGroupName: 'packt'
    location: 'East US'
    templateLocation: 'Linked artifact'
    csmFile: 'azure/azuredeploy.json'
    deploymentMode: 'Incremental'
```

> **Important note**
>
> The `Complete` deployment mode must be used with caution, and ensure you have a rigorous process where resources are only being created via templates. Otherwise, you could have destructive results that you were not expecting, such as services or applications not working anymore or loss of data.

Once the pipeline has been executed, you can validate the resources in the Azure portal:

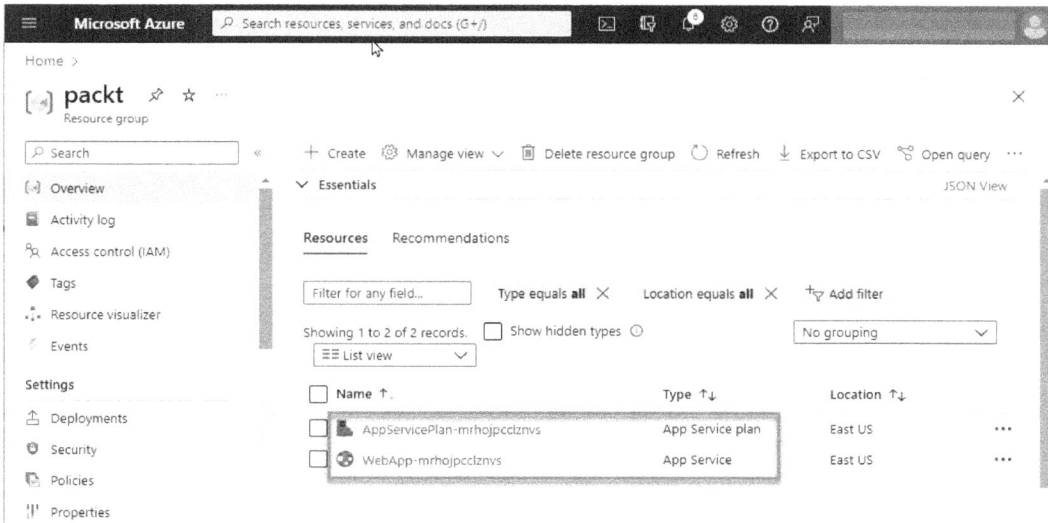

Figure 8.10 – Azure resources deployed via Azure Pipelines

If you want to learn more, head to `https://aka.ms/arm-syntax`. Now let's see how we can do something similar in AWS.

Working with AWS CloudFormation

AWS Cloud Formation is a service that allows you to define a template that describes a set of resources to be deployed together using JSON or YAML syntax. The templates follow this structure in the JSON format:

```
{
  "AWSTemplateFormatVersion" : "version date",
  "Description" : "JSON string",
  "Metadata" : { template metadata },
  "Parameters" : { set of parameters },
  "Rules" : { set of rules },
  "Mappings" : { set of mappings },
  "Conditions" : { set of conditions },
  "Transform" : { set of transforms },
  "Resources" : { set of resources },
  "Outputs" : { set of outputs }
}
```

Deploying with AWS CloudFormation comes down to the following steps:

1. Creating an IAM user with the AWS CLI
2. Creating a service connection to AWS
3. Creating an AWS CloudFormation template
4. Validating the AWS CloudFormation template
5. Deploying the AWS CloudFormation template

Let's start with discussing how to create an IAM user.

Creating an IAM user with the AWS CLI

An **IAM user** is a user defined in the **Identity and Access Management (IAM)** service in AWS and provides fine-grained access control to create, update, or delete resources in AWS, but is also used to grant/deny permissions to interact with other services.

> **Important note**
> This section assumes you have already configured your AWS credentials using the `aws`
> `configure` command.

Execute the following commands:

```
aws iam create-group --group-name resources-admin
aws iam attach-group-policy --group-name resources-admin --policy-arn
arn:aws:iam::aws:policy/AdministratorAccess
aws iam create-user --user-name azure-pipelines
aws iam add-user-to-group --group-name resources-admin --user-name
azure-pipelines
aws iam create-access-key --user-name azure-pipelines
```

These commands will do the following:

1. Create a user group called `resources-admin`
2. Attach a security policy to the user group
3. Create a user called `azure-pipelines`
4. Add the `azure-pipelines` user to the `resources-admin` user group
5. Create access keys for the `azure-pipelines` user

After executing these commands, you should have an output like that shown in the following screenshot:

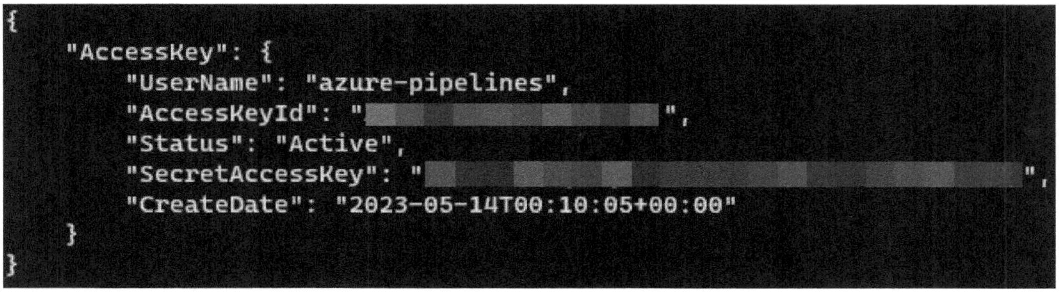

Figure 8.11 – AWS access keys for the azure-pipelines user

The `AccessKeyId` and `SecretAccessKey` values will be needed in the next step. Make sure to save them in a secure place as they provide programmatic access to AWS.

> **Important**
>
> The `arn:aws:iam::aws:policy/AdministratorAccess` policy used in this step is very permissive. It provides the highest level of access to the AWS console. This is not recommended for your environments. Instead, you should always provide access following a least-privilege approach and add more permissions when needed.

Let's now see how we create a service connection from Azure Pipelines to AWS.

Creating a service connection to AWS

As seen in previous chapters, integrating with external services in Azure Pipelines requires a service connection, which is done within **Project Settings | Pipelines | Service connections**, where you will click on the **New service connection** button and select the **AWS** option:

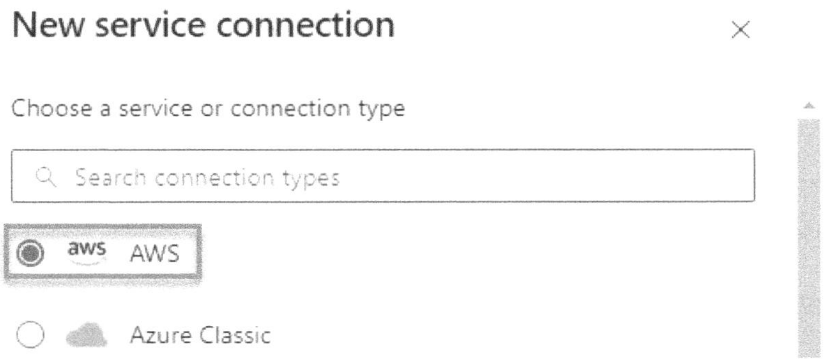

Figure 8.12 – AWS service connection type

The next step allows you to enter the following details:

- **Access Key ID**: Use the `AccessKeyId` value from the previous step
- **Secret Access Key**: Use the `SecretAccessKey` value from the previous step
- **Service connection name**: Use `aws-packt`

Now that we have a service connection, let's proceed to define the AWS CloudFormation template.

Creating an AWS CloudFormation template

You can create these templates from scratch following the reference documentation. Start from a sample template or use AWS CloudFormation Designer, a graphical tool that helps you create, visualize, and modify a template without worrying about all the formatting aspects. To learn more, go to `https://docs.aws.amazon.com/AWSCloudFormation/latest/UserGuide/template-guide.html`.

In this chapter, we will use a modified version of an existing sample that deploys a virtual machine using the **Elastic Compute Cloud** (**EC2**) service. You can find the template at `https://github.com/PacktPublishing/Implementing-CI-CD-Using-Azure-Pipelines/blob/main/ch08/aws/template.json`:

- An EC2 key pair – a set of security credentials used to connect to Linux instances
- An EC2 instance, which depends on the key pair and uses an Amazon Linux OS base image

Now let's see how we can validate the template.

Validating the AWS CloudFormation template

In your build or CI pipelines, you should consider validating your templates to ensure the format is correct. For this purpose, the AWS CLI task is available:

```
# AWS Cloud Formation Validation
trigger:
- main
pool:
  vmImage: ubuntu-latest
steps:
- task: AWSCLI@1
  inputs:
    awsCredentials: 'aws-packt'
    regionName: 'us-east-1'
    awsCommand: 'cloudformation'
```

```
awsSubCommand: 'validate-template'
awsArguments: '--template-body file://template.json'
```

Let's break down the parameters in the code:

- `awsCredentials` is a reference to the service connection.

- `regionName` is any of the available AWS Regions. Typically, you want to set this to the same region where the template would be deployed, to ensure it validates correctly based on the availability of services in that region.

- `awsCommand` is the top-level command in the AWS CLI that provides AWS CloudFormation operations.

- `awsSubcommand` indicates that you want to perform a template validation.

- `awsArguments` includes the options needed for the template validation to be completed. In this case, since the file was placed in the root folder of the repository, the template body is provided by passing the `file:` operator to read the contents of the file and perform the validation.

Once executed, the validation should show a message like that shown in the following screenshot:

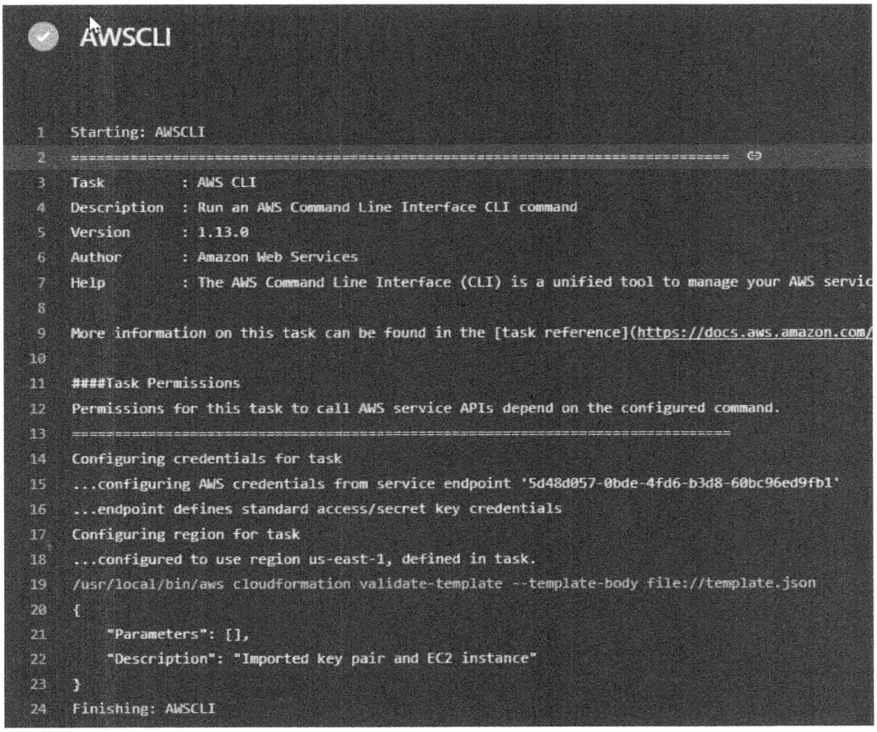

Figure 8.13 – AWS CloudFormation template successful validation

If there were an error, you would see a different message and task failure, as shown in the following screenshot, where a resource type was purposely mistyped:

```
An error occurred (ValidationError) when calling the ValidateTemplate
operation: Template format error: Unrecognized resource types:
[AWS::EC2::KeyXXXPair]
```

Now that we know how to validate the template, let's see how to deploy it.

Deploying the AWS CloudFormation template

Deploying the AWS CloudFormation template is known as creating an **AWS CloudFormation stack**, which is a service in AWS that allows you to group all resources in the template logically and has some additional benefits such as tracking drift, implementing a rollback strategy in case of errors, and the ability to delete the stack and all related resources.

In Azure Pipelines, you have two options to deploy the template:

- An AWS CLI task with the `aws cloudformation create-stack` command
- Using the `Cloud Formation Create or Update Stack` task, as shown in this code block:

```
# AWS Cloud Formation Deployment
trigger:
- main
pool:
  vmImage: ubuntu-latest
steps:
- task: CloudFormationCreateOrUpdateStack@1
  inputs:
    awsCredentials: 'aws-packt'
    regionName: 'us-east-1'
    stackName: 'packt'
    templateSource: 'file'
    templateFile: 'template.json'
    capabilityIAM: false
    capabilityNamedIAM: false
    onFailure: 'DELETE'
```

Let's break the preceding code down:

- `awsCredentials` is a reference to the service connection.
- `regionName` is any of the AWS regions available.
- `stackName` is a name to identify this stack in the AWS console and must be unique.

- `templateSource` in this case is `file`; however, it could also be a `url` to a template, perhaps outside of Azure Pipelines; `s3`, which refers to a storage service in AWS, and you would provide a bucket and object key; or `usePrevious`, which indicates you want to use the template in an existing stack.

- `templateFile` is used to provide the location of the file with the template and is only needed because `file` has been set to `templateSource`.

- The `capibilityIAM` and `capabilityNamedIAM` parameters are set to `false`. These are additional properties needed for some type of deployment where IAM changes would be applied. In the case of the template used in this example, they are not required.

- Finally, the `onFailure` property indicates what to do with the stack if something goes wrong. With the `DELETE` value, it would be deleted and any resources that might have been deployed successfully would be deleted as well. The `DO_NOTHING` value would simply stop applying the template and you would be able to see in the AWS console what had happened so far. Lastly, the `ROLLBACK` value, which is the default, would revert any changes made prior to applying the template.

With the deployment completed successfully, you should be able to see the status in the AWS console, as shown in the following screenshot:

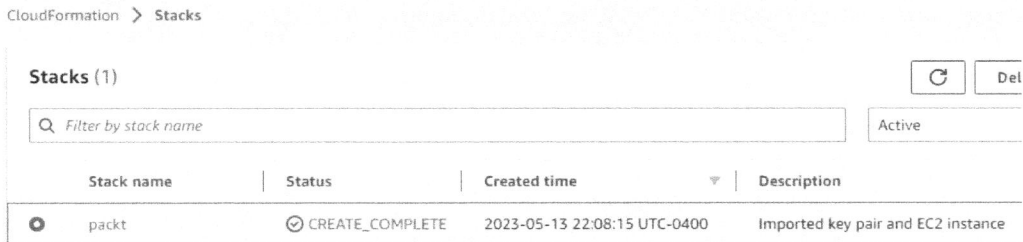

Figure 8.14 – AWS CloudFormation stack deployed successfully

You will also be able to see the resources deployed as part of the CloudFormation stack, as shown in the following screenshot:

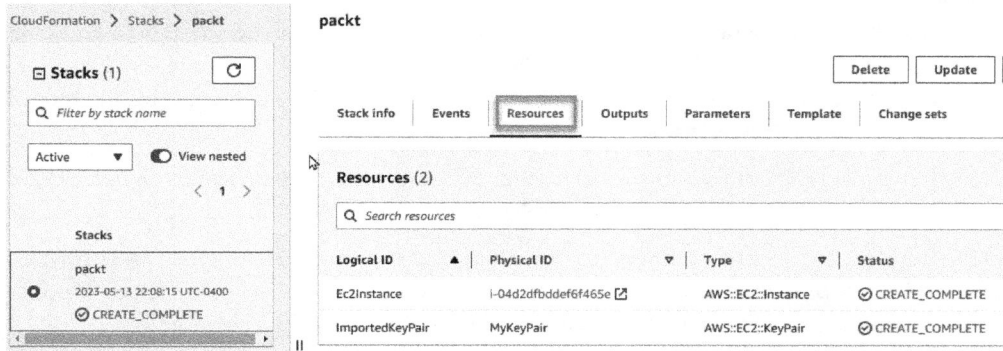

Figure 8.15 – AWS CloudFormation stack resources

So far, we have learned about Azure and AWS cloud platform-specific IaC capabilities; however, these only work for each of these platforms. If you want to do IaC in a generic way and aim for multiple types of targets – say, on-premises and other cloud providers, there are other tools on the market that can accomplish this.

Some other IaC tools are Ansible, Chef, Pulumi, Puppet, SaltStack, and Terraform, to name a few. In this chapter, we will focus on **Terraform** because it is one of the most popular choices among the open source community due to its versatility and declarative nature. Let's next learn how to use Terraform for IaC.

Working with Terraform

Let's first learn about how Terraform works and then we will walk through using it within Azure Pipelines.

How does Terraform work?

Terraform is a tool that allows you to write IaC and define resources for both cloud and on-premises resources using a domain-specific language. It uses providers as a means to encapsulate the resource definition for supported targets.

The following diagram depicts the high-level architecture of Terraform:

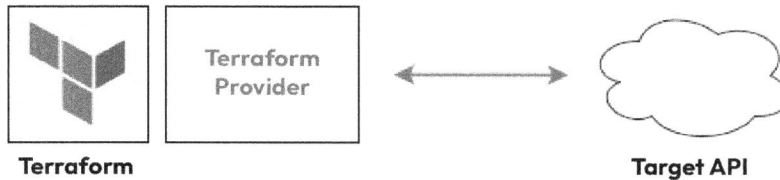

Figure 8.16 – Terraform architecture

It works by way of the following three steps:

1. **Write**: You define the resources in templates to deploy all the resources you need, across the targets required. There could be more than one.

2. **Plan**: Terraform creates an execution plan to determine the changes that need to be made to match the definition, calculating the sequential order of operations and understanding any resource dependencies. This could mean creating resources, updating them, or destroying them.

3. **Apply**: Once you agree with the plan, Terraform performs the necessary operations in the sequence calculated by the plan.

Part of the process Terraform follows involves calculating the state of the target in comparison to the plan. This is called the **Terraform state file**. This file will contain details about the resources, their metadata in the target destination, and their dependencies.

Depending on the edition of Terraform you are using, you might be responsible for managing the location of this file or use built-in features to manage the state file for you.

Managing the location of the state file is critical in Terraform. It can be stored locally or remotely via a **backend** configuration. When you are getting started, you will likely use the **local backend**; however, as you evolve your configuration to support multiple environments, you will switch to a **remote backend**. To read more about remote backends, go to `https://developer.hashicorp.com/terraform/language/settings/backends/remote`.

> **Important note**
>
> Terraform works in absolutes, meaning it expects to control every resource defined in the templates you create by comparing them with the current state file every time. This means that no changes should be made without the use of Terraform in the environment. Otherwise, these changes will be lost the next time you run the `terraform apply` command.

Terraform comes in three editions:

- **Open source**: Free, downloadable, and flexible to use with your existing source control and CI/CD tools
- **Cloud**: A SaaS application that allows you to run it in a stable and remote environment, with secure storage of the state files and secrets
- **Enterprise**: Allows you to set up a private Terraform Cloud instance or self-hosted distribution with customizable resource limits and tighten security.

Let's now look at how to create a simple Terraform template.

Creating a Terraform template

Terraform templates are usually written in what is called configuration syntax, which is a pseudo-JSON format. All files have the `.tf` extension and typically you will have the following files:

- `providers.tf`: Defines the base configuration and providers to use
- `main.tf`: The entry point of your template – say, your starting point
- `variables.tf`: Defines values to be used throughout the templates that can be overridden when planning/applying the configuration
- `outputs.tf`: Values to extract from the resources deployed

In this chapter, we will use a simple set of templates targeting the Azure cloud platform, creating a resource group, and using the `azurerm` backend. See the templates at `https://github.com/PacktPublishing/Implementing-CI-CD-Using-Azure-Pipelines/tree/main/ch08/terraform`.

For more details on the Terraform configuration language, head to `https://developer.hashicorp.com/terraform/language`. For tutorials and training material, head to `https://developer.hashicorp.com/tutorials/library?product=terraform`.

Now that we have learned the Terraform basics for creating templates, let's put together a pipeline to validate them.

Managing the Terraform state file

Before we can put together a build pipeline, we need to ensure that the **state file** is managed accordingly – in this case, remotely. Since our template will target Azure, we will need to set up the corresponding backend for it.

For this, you can execute the following Azure CLI commands, once logged in:

```
az group create --name tfstate --location eastus
az storage account create --name tfstate --resource-group tfstate
--location eastus --sku Standard_LRS
az storage container create --name tfstate --account-name tfstate
```

These commands perform the following operations:

- Create a resource group in Azure named `tfstate` in the `eastus` region
- Create a storage account in Azure named `tfstate` in the `tfstate` resource group in the `eastus` region with the `Standard_LRS` pricing tier
- Create a blob container named `tfstate` in the `tfstate` storage account

> **Important**
>
> The storage account name in the previous commands must be unique globally, so you will have to adjust the name in these commands and any following sections for everything to work correctly.

Now that we have set up the state file management, let's put together a pipeline to validate the template.

Validating a Terraform template

In your build or CI pipelines, you should consider validating your templates to ensure the format is correct. For this purpose, tasks are available in the Terraform Marketplace extension:

```
# Terraform pipeline
trigger:
- main
pool:
  vmImage: ubuntu-latest
steps:
- task: TerraformInstaller@0
  displayName: 'install'
  inputs:
    terraformVersion: 'latest'
- task: TerraformTaskV4@4
  displayName: 'init'
  inputs:
    provider: 'azurerm'
    command: 'init'
    backendServiceArm: 'azure-packt-rg'
    backendAzureRmResourceGroupName: 'tfstate'
```

```
      backendAzureRmStorageAccountName: 'tfstate'
      backendAzureRmContainerName: 'tfstate'
      backendAzureRmKey: 'terraform.tfstate'
- task: TerraformTaskV4@4
  displayName: 'validate'
  inputs:
    provider: 'azurerm'
    command: 'validate'
```

Now let's break this code down:

- The `TerraformInstaller@0` task installs the Terraform CLI in the agent if necessary. This is required if you need to ensure a specific version of the tool is required for your templates. Microsoft-hosted agents typically have a version of Terraform installed, but if you want to use an older or newer version, this task will allow you to use the version of Terraform that you need.

- The `TerraformTask@4` task allows you to run any of the Terraform CLI commands. For the first task with `displayName: init`, it will correspondingly execute the `terraform init` command with the following parameters:

  - `backendServiceArm` indicating the name of the ARM service connection to use

  - `backendAzureRmResourceGroupName` indicating the resource group in Azure where the state file will be stored

  - `backendAzureRmStorageAccountName` with the name of the Azure storage account where the state file will be stored

  - `backendAzureRmContainerName` with the blob container name in the Azure storage account where the state file will be stored

  - `backendAzureRmKey` indicating the name of the state file

- Finally, the last task with `displayName: validate` will execute the `terraform validate` command.

Once the pipeline runs, the validation should complete successfully and you should see a message similar to the following for the validation task:

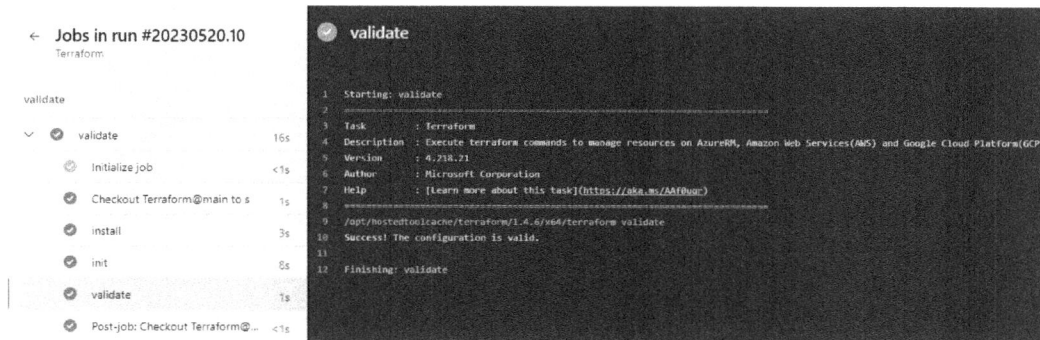

Figure 8.17 – Successful Terraform template validation

Now that we've learned how to validate the template, let's proceed to deploy resources with Terraform.

Deploying a Terraform template

As mentioned previously, deploying resources with Terraform is a two-step process, requiring you to execute the `plan` and `apply` commands:

```
# Terraform pipeline
trigger:
- main
pool:
  vmImage: ubuntu-latest
steps:
- task: TerraformInstaller@0
  displayName: 'install'
  inputs:
    terraformVersion: 'latest'
- task: TerraformTaskV4@4
  displayName: 'init'
  inputs:
    provider: 'azurerm'
    command: 'init'
    backendServiceArm: 'azure-packt-rg'
    backendAzureRmResourceGroupName: 'tfstate'
    backendAzureRmStorageAccountName: 'tfstate'
    backendAzureRmContainerName: 'tfstate'
    backendAzureRmKey: 'terraform.tfstate'
- task: TerraformTaskV4@4
  displayName: 'plan'
  inputs:
```

```
    provider: 'azurerm'
    command: 'plan'
    environmentServiceNameAzureRM: 'azure-packt-rg'
- task: TerraformTaskV4@4
  displayName: 'apply'
  inputs:
    provider: 'azurerm'
    command: 'apply'
    environmentServiceNameAzureRM: 'azure-packt-rg'
```

Let's break this code down:

- The `TerraformInstaller@0` task installs the Terraform CLI in the agent if necessary
- The `TerraformTask@4` task with `displayName: init` initializes Terraform as explained in the previous section
- The `TerraformTask@4` task with `displayName: plan` executes the `terraform plan` command, performing the comparison necessary and updating the state file accordingly with changes to be made
- The `TerraformTask@4` task with `displayName: apply` executes the `terraform apply` command and performs the changes necessary in Azure

We covered a lot of ground in this chapter. Let's finish off with a summary of what we learned.

Summary

In this chapter, we learned about different tools available to create, update, and delete resources on the Microsoft Azure and AWS cloud platforms.

We learned how to create, validate, and deploy ARM templates, the role of service principals in Azure, and the security considerations of deploying with automation.

We also learned about AWS CloudFormation templates and stacks, and how to create them and update them from Azure Pipelines. At the same time, we learned how AWS has a similar security model to Azure and about the security implications of credentials.

Finally, we learned about Terraform as an abstraction language to define IaC for on-premises and cloud platforms such as Azure and AWS, and how to validate templates in Azure Pipelines and deploy resources with it.

Regardless of which IaC tool you choose to use, they are important because they will allow you to do the following:

- Have a repeatable and immutable process for deployments
- Accelerate deployments
- Reduce or eliminate errors while deploying with Azure Pipelines
- Recover from issues easily
- Spend more time working on your application code

In the next chapter, we will be putting everything we have learned together for end-to-end pipeline building and packaging applications to deploy them on different Microsoft Azure cloud platform services.

Part 3: CI/CD for Real-World Scenarios

Finally, in the last part of this book, we will compile everything we've learned so far to create end-to-end scenarios typically found in the real world, using cloud platforms, and we will walk through some best practices.

This part has the following chapters:

- *Chapter 9, Implementing CI/CD for Azure Services*
- *Chapter 10, Implementing CI/CD for AWS*
- *Chapter 11, Automating CI/CD for Cross-Mobile Applications by Using Flutter*
- *Chapter 12, Navigating Common Pitfalls and Future Trends in Azure Pipelines*

Implementing CI/CD
for Azure Services

In this chapter, you are going to put everything you have learned so far into an end-to-end solution, deploying a group of applications and promoting it from a test environment to a production environment. From the simplest to the most complex solution architecture, this chapter showcases the flexibility of Azure Pipelines to handle the provisioning, configuration, and deployment of applications in Azure, no matter the different services involved and the programming languages used in your applications.

The following topics will be covered:

- Introducing the solution architecture
- Building and packaging applications and **infastruture as code** (**IaC**)
- Creating environments
- Deploying a Python catalog service to **Azure Kubernetes Service** (**AKS**)
- Deploying a Node.js cart service to **Azure Container Apps** (**ACA**)
- Deploying a .NET checkout service to **Azure Container Instances** (**ACI**)
- Deploying an Angular frontend app to **Azure App Service** (**AAS**)
- Approving environment deployments

Let's first take care of the technical requirements for this chapter.

Technical requirements

You will need to have handy the following URL to the GitHub repository, which will be the base for this chapter: `https://github.com/PacktPublishing/Implementing-CI-CD-Using-Azure-Pipelines/tree/main/ch09`.

Like in the previous chapter, you must have access to an Azure account to complete the steps in this chapter; if you don't have one, you can create a free one at `https://azure.microsoft.com/en-us/free/`.

That's it for the technical requirements; now, let's get started with this chapter.

Getting started

The first thing you must do is import the sample repository; let's get to it.

Importing the sample repository

You will need to import the application and IaC sources from GitHub for this book, to be able to complete the end-to-end pipelines in this chapter and the next.

You can do this from the **Azure Repos | Files** section in the navigation menu, clicking on the repository dropdown on the top part of the screen and then clicking on the **Import repository** option, as shown in the following screenshot:

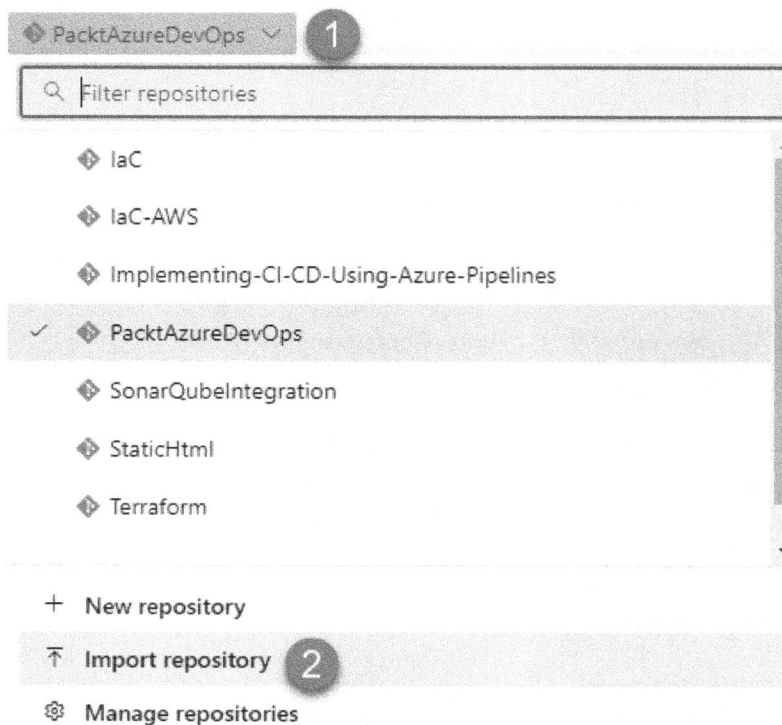

Figure 9.1 – Import repository

Enter `https://github.com/PacktPublishing/Implementing-CI-CD-Using-Azure-Pipelines.git` in the **Clone URL** field and `Implementing-CI-CD-Using-Azure-Pipeline` in the **Name** field, and then hit the **Import** button:

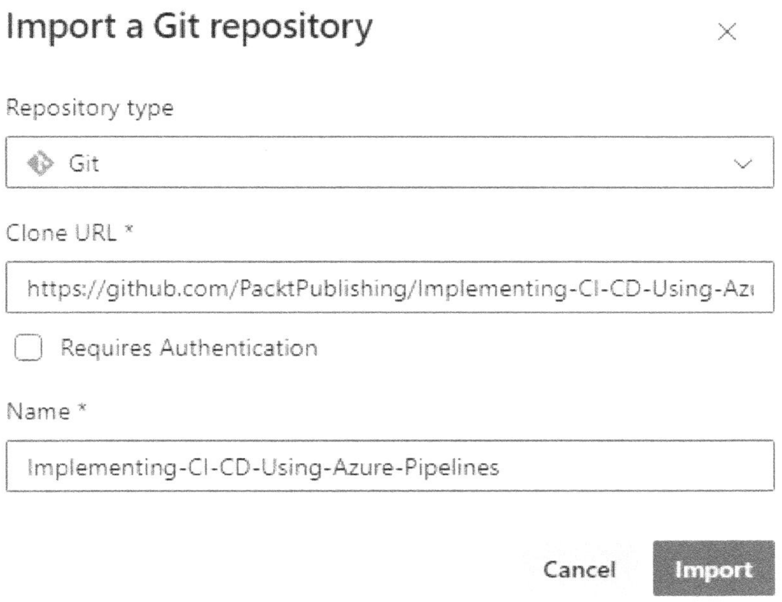

Import a Git repository   ✕

Repository type

◈ Git   ∨

Clone URL *

https://github.com/PacktPublishing/Implementing-CI-CD-Using-Azu

☐ Requires Authentication

Name *

Implementing-CI-CD-Using-Azure-Pipelines

Cancel   **Import**

Figure 9.2 – Importing the sample repository

After a few minutes, the process will be complete, and then you will be able to browse all the code in the repository. Everything you need will be in the `e2e` directory.

With the repository imported, let's understand the sample architecture.

Introducing the solution architecture

For our sample architecture, you will use a fictitious Packt store made up of four different applications, which represents a complex distributed architecture where teams working with different programming languages can use different Azure platform services to deliver their capabilities:

- An Angular frontend application, the user interface of the store
- A Python product catalog service, implemented as a REST API
- A Node.js shopping cart service, implemented as a REST API
- An ASP.NET checkout service, implemented as a REST API

The following solution diagram depicts how an environment for the web store looks, where each of the applications independently runs in a different Azure service:

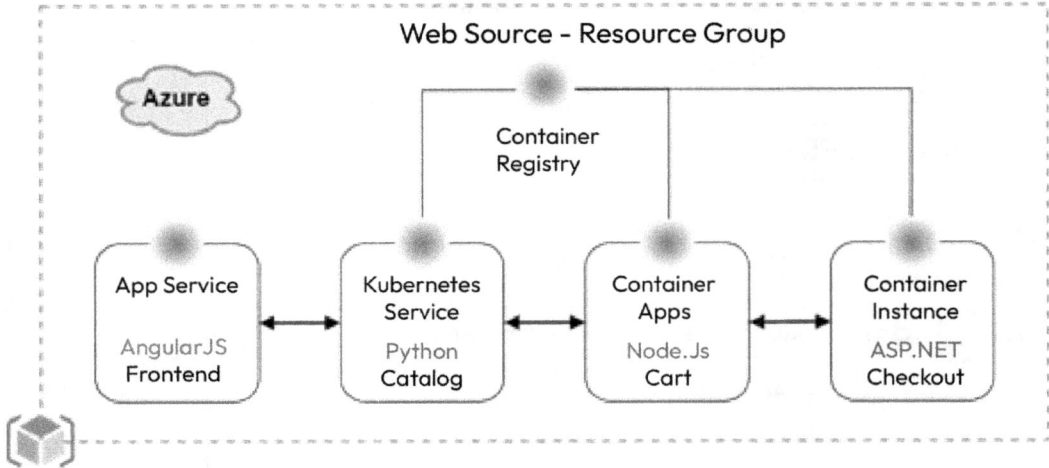

Figure 9.3 – The solution diagram

Later in the chapter, you will implement the Azure Pipeline with the following steps for each application:

1. Build and package the application and its corresponding IaC.

2. Deploy them to a test environment.

3. Deploy them to a production environment, including a manual approval check.

The following diagram depicts the CI/CD process:

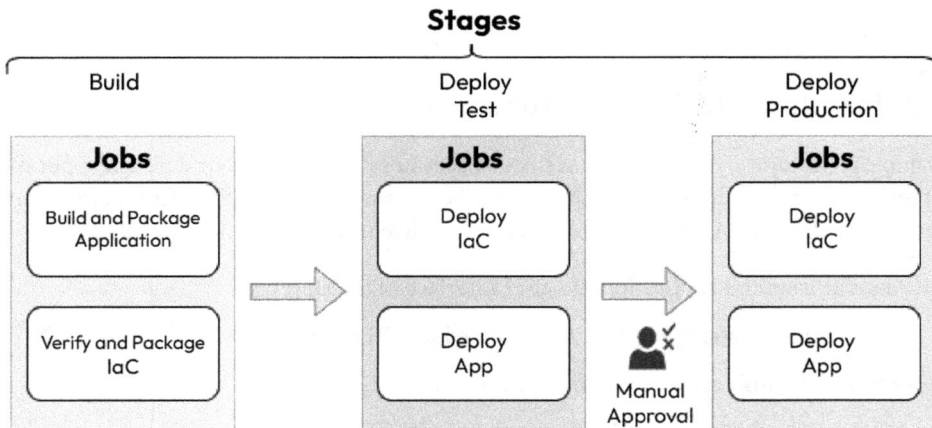

Figure 9.4 – The CI/CD end-to-end process

> **Important note**
>
> During this chapter, no details about the code in the applications will be covered, as that is not relevant to CI/CD. We will focus instead on the Azure Pipelines details needed to make the CI/CD process work.

To implement the CI/CD process, you will be take advantage of multi-stage pipelines with environments and templates:

- **Stages** allow us to encapsulate all the jobs that need to happen together in a logical way and control dependencies, but they also provide the ability to execute jobs in parallel; this will help us reduce the total time when needed

- **Environments** are linked to jobs and allow us to add additional controls, such as manual approval in this case, to ensure that the deployment proceeds to production only when a human intervenes and approves it

- **Templates** were introduced in *Chapter 1*; here, you will put them into practice to demonstrate how to use them, to build modular and reusable pipelines

Let's look at the following pipeline definition; create this file in `ch09/azure/azure-pipeline.yml`, inside the `Implementing-CI-CD-Using-Azure-Pipeline` repository you imported:

```
# Multi-Stage pipeline
trigger:
- main
pool:
  vmImage: ubuntu-latest
stages:
- stage: build
  displayName: Build
  jobs:
  - template: build-apps.yml
  - template: build-iac.yml
- stage: deployTest
  displayName: Deploy Test
  dependsOn: build
  jobs:
  - template: deploy.yml
    parameters:
      envName: test

 - stage: deployProduction
   displayName: Deploy Production
   dependsOn: deployTest
```

```
jobs:
- template: deploy.yml
  parameters:
    envName: production
```

Let's break this code down:

- The `build` stage has no dependencies and includes jobs from templates in the `build-apps.yml` and `build-iac.yml` files. You will review these later in the *Building and packaging applications and IaC* section.

- The `deployTest` stage must wait for the build stage to complete and will run the jobs in the `deploy.yml` template, passing a parameter of `envName` with the `test` value to uniquely identify this environment.

- The `deployProduction` stage, in turn, waits for the `deployTest` stage to complete and uses the same `deploy.yml` template, passing a value of production for the `envName` parameter.

This pipeline definition demonstrates the flexibility of templates and the ability to break into smaller portions the work that needs to be done, which provides a way for teams to focus on different stages of the pipeline, based on their responsibilities.

Once the file is in the repository, add it as a new pipeline and rename it `E2E-Azure`. You will have to add some security configuration for everything to ultimately work.

If you have not renamed a pipeline before, click on the sub-menu on the far-right side of the **Recently run pipelines** screen, as shown in the following screenshot, and rename it:

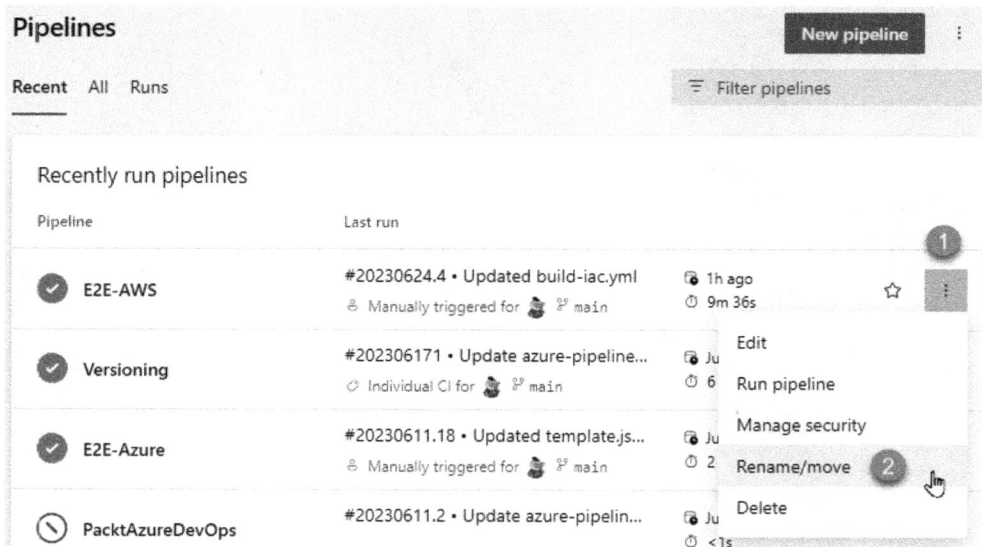

Figure 9.5 – Renaming a pipeline

> **A pro tip about templates**
>
> When using `job.template` to reference a template defined in another file, make sure to create that file with at least one step defined in it before referencing it from the parent template. An easy way to accomplish this is to use a `script` task that runs an `echo` command, such as `echo hello`.

Let's now move on to the build stage.

Building and packaging applications and IaC

The applications in this solution are all container-enabled, a standard packaging mechanism that includes all operating system dependencies to allow them to run in many different hosting environments, making them extremely lightweight and portable.

For simplicity, the repository includes a `docker-compose.yml` file, which facilitates working with applications made of multiple services that must run at the same time.

This file defines the services and the location of their corresponding `Dockerfile`, the file that defines how the container must be built, and several other things, such as the ports or environment variables needed for the container to run.

In this chapter, the `SUB_ID` placeholder is the ID of the Azure subscription you have access to; make sure to replace it when appropriate.

Before you can proceed, you must have an **Azure Container Registry** available for the pipeline to store the container images. You can create one easily with the following Azure CLI command:

```
az acr create -n packtadocicd -g packt -sku Standard -l eastus -admin-enabled
```

We will also need to ensure that the service principal we created previously in the *Creating a service connection to AWS* section of *Chapter 8* has permissions to manage user access to this service; you can do this with the following commands in the PowerShell window:

```
$id=az ad sp list -display-name azure-pipelines -query "[].id" -o tsv
az role assignment create -assignee-object-id $id -scope /subscriptions/SUB_ID/resourceGroups/packt/providers/Microsoft.ContainerRegistry/registries/packtadocicd --role 'User Access Administrator'
```

With the registry in place, you can now move on to create the `build-apps.yml` file, which will be used to build the application containers and push them to the Azure Container Registry.

To build and push the containers, you can use the Docker Compose task. However, this must be done as a two-step process; you must build the images first and then push them. To make the task easier to read, let's first look at the build portion in the following YAML code:

```yaml
parameters:
- name: azureSubscription
  type: string
  default: 'azure-packt-rg'
- name: azureContainerRegistry
  type: string
  default: '{"loginServer":"packtadocicd.azurecr.io", "id" : "/
subscriptions/SUB_ID/resourceGroups/packt/providers/Microsoft.
ContainerRegistry/registries/packtadocicd"}'
jobs:
- job: BuildAndPushContainers
  displayName: Build and Push Containers
  steps:
  - task: DockerCompose@0
    displayName: 'Build Containers'
    inputs:
      containerregistrytype: 'Azure Container Registry'
      azureSubscription: ${{parameters.azureSubscription}}
      azureContainerRegistry: ${{parameters.azureContainerRegistry}}
      dockerComposeFile: 'docker-compose.yml'
      projectName: 'packt-store'
      action: 'Build services'
      additionalImageTags: '$(Build.BuildNumber)'
      includeLatestTag: true
```

Let's break this code down:

- The `azureSubscription` parameter is a reference to the Azure Resource Manager Service connection created in the previous chapter.

- The `azureContainerRegistry` parameter is a little tricky; it is a JSON document that includes the `loginServer` and `id` properties related to the Container Registry resource in Azure.

- The Docker Compose task with the `'Build Containers'` `displayName` uses the `docker-compose.yml` file to build container images locally in the agent, as indicated by the `action` property. Note the use of `additionalImageTags`, where you provide a predefined variable, `$(Build.BuildNumber)`, and set to `true` the `includeLatestTag` property; we will elaborate on this in the *Understanding container image tags* section.

With the images built, the next step is to push them to the registry, which is done with the same task, with just a change to the `action` property, as shown in the following code snippet:

```
- task: DockerCompose@0
  displayName: 'Push Containers'
  inputs:
    containerregistrytype: 'Azure Container Registry'
    azureSubscription: ${{parameters.azureSubscription}}
    azureContainerRegistry: ${{parameters.azureContainerRegistry}}
    dockerComposeFile: 'docker-compose.yml'
    projectName: 'packt-store'
    action: 'Push services'
    additionalImageTags: '$(Build.BuildNumber)'
    includeLatestTag: true
```

The Docker Compose task with `displayName` `'Push Containers'` uses the `docker-compose.yml` file to push the previously built container images to the Container Registry in Azure as indicated by the `action` property.

Remember that the two portions of YAML presented here are part of the `build-apps.yaml` file.

Now that we have covered how to build and push container images, let's take a break to discuss how container image tags work and why they are important.

Understanding container image tags

Building a container is like compiling an application and packaging all its files into a ZIP archive that you can then use for deployment, along with all the OS dependencies needed for the application to run.

However, the result is called a container image and it is typically a complex artifact made up of multiple layers stored in a registry and is not manageable via the filesystem. For this reason, just like you would name a ZIP file based on a versioning convention to track when the artifact was generated, when working with containers, it is important to tag them.

The `latest` tag mentioned previously is a convention in the container world that allows you to retrieve the newest image available without specifying a specific tag. This is very helpful during development cycles for experimentation purposes.

> **Important note**
>
> Always tag your containers with a specific version number and deploy that version number across all your environments for proper traceability. The `latest` tag is only a convenience to easily pull the newest version of a container image and it should never be used for environment deployments, because it can be a reference to different builds depending on the date and time that you pull it.

Now that you understand the importance of container image tags, let's see what you can use within Azure Pipelines to name them.

Understanding your pipeline build number

The $(Build.BuildNumber) predefined variable is a convenient way to get a unique label in every pipeline run to version your artifacts, and its default value is a timestamp and revision number with the format YYYYMMDD.R, where YYYY is the current year, MM is the month, DD is the day, and R is a sequential automatically incremented number.

If you don't set your build name explicitly, it will use the following default format for your YAML pipelines:

```
name: $(Date:yyyyMMdd).$(Rev:r)
```

This special notation will use the current date in the given format, automatically increase the number generated by the $(Rev:r) token, and reset it to 1 if the portion of text before it is changed.

Most organizations prefer to use semantic versioning for artifacts or APIs, which follows a MAJOR.MINOR.PATCH format, like 1.0.1:

- MAJOR changes indicate incompatible API changes
- MINOR changes indicate that functionality is added with backward compatibility
- PATCH changes indicate a bug fix without impact on functionality

If you need to use semantic versioning in your pipelines, an easy way to implement this is by adding name at the very top of your YAML file, as follows:

```
name: 1.0.$(Rev:r)
```

Notice that in this case, you are responsible for increasing the MAJOR and MINOR portions of the name based on your code changes.

> **Important note**
> Always consider the implications of your pipeline build number, its format, and where you will be using it. This value can have adverse effects depending on where you use it.

Now that you have the container images available, let's talk a bit about Helm, a tool you will be using to deploy applications in Kubernetes environments.

Understanding Helm

Helm is a package manager for Kubernetes. Typically, you take advantage of it to deploy third-party or open source applications into your Kubernetes clusters. The packages created with Helm are referred to as **Helm charts**.

Helm is also extremely useful for packaging your own applications, since they will likely have more than one manifest needed to configure all required components in Kubernetes, and Helm provides facilities to override parameters with ease.

For example, a simple Helm chart will contain the following files:

```
        .helmignore
        Chart.yaml
        values.yaml

     ──charts
     ──templates
            deployment.yaml
            hpa.yaml
            ingress.yaml
            NOTES.txt
            service.yaml
            serviceaccount.yaml
            _helpers.tpl
```

Figure 9.6 – Basic Helm chart contents

If you want to learn more about Helm, go to `https://helm.sh/`.

Validating Helm charts is not a trivial task; several options are available for this. Helm provides a basic `lint` command to accomplish this task, but it covers only basic format issues. In this book, you are using an open source tool called **kube-linter**, available as a Docker container. This will validate the YAML syntax of the Kubernetes manifests used to deploy the application and perform a series of best-practice checks. If you want to learn more about this tool, go to `https://docs.kubelinter.io/`.

Now that you have learned about Helm, let's work on the IaC.

Verifying and packaging IaC

You learned how to work with Azure Resource Manager templates in the previous chapter, so you need to validate the templates and publish them as artifacts to the pipeline.

To do this, you will create a `build-iac.yml` file in the repository and add the following seven segments to it (they have been separated in this section only to make it easier to read):

- **The parameters segment**: These parameters will allow us to easily replace values when called from the main `azure-pipeline.yaml` file:

```
parameters:
- name: azureSubscription
  type: string
  default: 'azure-packt-rg'
- name: resourceGroupName
  type: string
  default: 'packt'
- name: location
  type: string
  default: 'East US'
```

- **The jobs segment**: This segment groups all the subsequent segments that include only tasks:

```
jobs:
- job: VerifyAndPackageIaC
  displayName: Verify and Package IaC
  steps:
```

- **IaC catalog tasks segment**: This can be written as follows:

```
    - task: AzureResourceManagerTemplateDeployment@3
      displayName: 'Validate Catalog Template'
      inputs:
        deploymentScope: 'Resource Group'
        azureResourceManagerConnection: ${{parameters.
azureSubscription}}
        resourceGroupName: ${{parameters.resourceGroupName}}
        templateLocation: 'Linked artifact'
        csmFile: 'e2e/iac/azure/catalog/template.json'
        deploymentMode: 'Validation'
        location: ${{parameters.location}}
    - task: PublishPipelineArtifact@1
      displayName: 'Publish Catalog Artifacts'
      inputs:
        targetPath: 'e2e/iac/azure/catalog'
```

```
    artifact: catalog-iac
    publishLocation: 'pipeline'
```

Let's break it down:

- The `AzureResourceManagerTemplateDeployment@3` task is used to validate the ARM templates for the catalog application

- The `PublishPipelineArtifact@1` task is then used to publish the artifacts to be used for deployment

- **Catalog helm chart segment**: You can write this block as follows:

```
- script: |
    docker run --rm -v $(pwd):/manifests stackrox/kube-
linter lint /manifests --config /manifests/.kube-linter.yml
    displayName: 'Lint Catalog Helm Chart'
    workingDirectory: e2e/iac/helm-charts/catalog
- task: HelmInstaller@1
  displayName: 'Install Helm'
- task: HelmDeploy@0
  displayName: 'Package Catalog Helm Chart'
  inputs:
    command: package
    chartPath: e2e/iac/helm-charts/catalog
    destination: $(Build.ArtifactStagingDirectory)
- task: PublishPipelineArtifact@1
  displayName: 'Publish Catalog Helm Chart'
  inputs:
    targetPath: $(Build.ArtifactStagingDirectory)
    artifact: catalog-helm-chart
    publishLocation: 'pipeline'
```

Let's break it down:

- The script task with `displayName` `'Lint Catalog Helm Chart'` performs a validation of the Helm chart

- The `HelmInstaller@1` task installs the Helm tool

- The `HelmDeploy@0` task is used to package the Helm chart

- **IaC cart tasks segment**: An example of this is as follows:

```
- task: AzureResourceManagerTemplateDeployment@3
  displayName: 'Validate Cart Template'
  inputs:
    deploymentScope: 'Resource Group'
```

```
        azureResourceManagerConnection: ${{parameters.
azureSubscription}}
        resourceGroupName: ${{parameters.resourceGroupName}}
        templateLocation: 'Linked artifact'
        csmFile: 'e2e/iac/azure/cart/template.json'
        deploymentMode: 'Validation'
        location: ${{parameters.location}}
    - task: PublishPipelineArtifact@1
      displayName: 'Publish Cart Artifacts'
      inputs:
        targetPath: 'e2e/iac/azure/cart'
        artifact: cart-iac
        publishLocation: 'pipeline'
```

Let's break it down:

* The `AzureResourceManagerTemplateDeployment@3` task is used to validate the ARM templates for the cart application

* The `PublishPipelineArtifact@1` task is then used to publish the artifacts to be used for deployment

* **IaC checkout tasks segment**: This section looks like the following:

```
    - task: AzureResourceManagerTemplateDeployment@3
      displayName: 'Validate Checkout Template'
      inputs:
        deploymentScope: 'Resource Group'
        azureResourceManagerConnection: ${{parameters.
azureSubscription}}
        resourceGroupName: ${{parameters.resourceGroupName}}
        templateLocation: 'Linked artifact'
        csmFile: 'e2e/iac/azure/checkout/template.json'
        deploymentMode: 'Validation'
        location: ${{parameters.location}}
    - task: PublishPipelineArtifact@1
      displayName: 'Publish Checkout Artifacts'
      inputs:
        targetPath: 'e2e/iac/azure/checkout'
        artifact: checkout-iac
        publishLocation: 'pipeline'
```

Let's break it down:

- The `AzureResourceManagerTemplateDeployment@3` task is used to validate the ARM templates for the checkout application

- The `PublishPipelineArtifact@1` task is then used to publish the artifacts to be used for deployment

- **IaC frontend tasks segment**: Here is a sample of this part of the code:

```
- task: AzureResourceManagerTemplateDeployment@3
  displayName: 'Validate Frontend Template'
  inputs:
    deploymentScope: 'Resource Group'
    azureResourceManagerConnection: ${{parameters.
azureSubscription}}
    resourceGroupName: ${{parameters.resourceGroupName}}
    templateLocation: 'Linked artifact'
    csmFile: 'e2e/iac/azure/frontend/template.json'
    deploymentMode: 'Validation'
    location: ${{parameters.location}}
    overrideParameters: '-catalogAppUrl catalogAppUrl.com
 -cartAppUrl cartAppUrl.com -checkoutAppUrl checkoutAppUrl.com'
- task: PublishPipelineArtifact@1
  displayName: 'Publish Frontend Artifacts'
  inputs:
    targetPath: 'e2e/iac/azure/frontend'
    artifact: frontend-iac
    publishLocation: 'pipeline'
```

For simplicity, let's break down what is happening:

- The `AzureResourceManagerTemplateDeployment@3` task is used to validate the ARM templates for the frontend application

- The `PublishPipelineArtifact@1` task is then used to publish the artifacts to be used for deployment

That brings us to the end of the `build-iac.yaml` file; make sure you keep all the segments together in the same file.

Now that you have all the artifacts ready, let's move on to create our environments.

Managing environments

In this section, you will learn about how to create environments and deploy to them.

Configuring environments

In this section, you will define the environments in Azure Pipelines, which will be logical representations of the deployment targets. This will allow us to add approval and checks to control how the pipeline advances from one stage to the next:

1. You start by clicking on the **Environments** option under **Pipelines** in the main menu on the left, as follows:

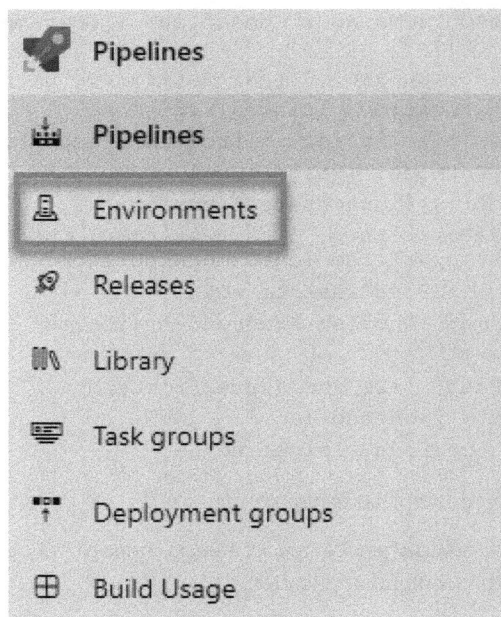

Figure 9.7 – Accessing the Environments option in the menu

2. If you have no environments, you will see a screen like the following; click on **Create environment**:

Create your first environment

Manage deployments, view resource status and get full end to end traceability

Create environment

Figure 9.8 – Creating your first environment

Otherwise, you will see a **New environment** option in the top-right part of the screen above your existing environments.

3. Once the pop-up screen shows up to create the new environment, enter `test` for **Name** and `Test Environment` for **Description**, leave the **Resource** option as **None**, and click on the **Create** button, as shown here:

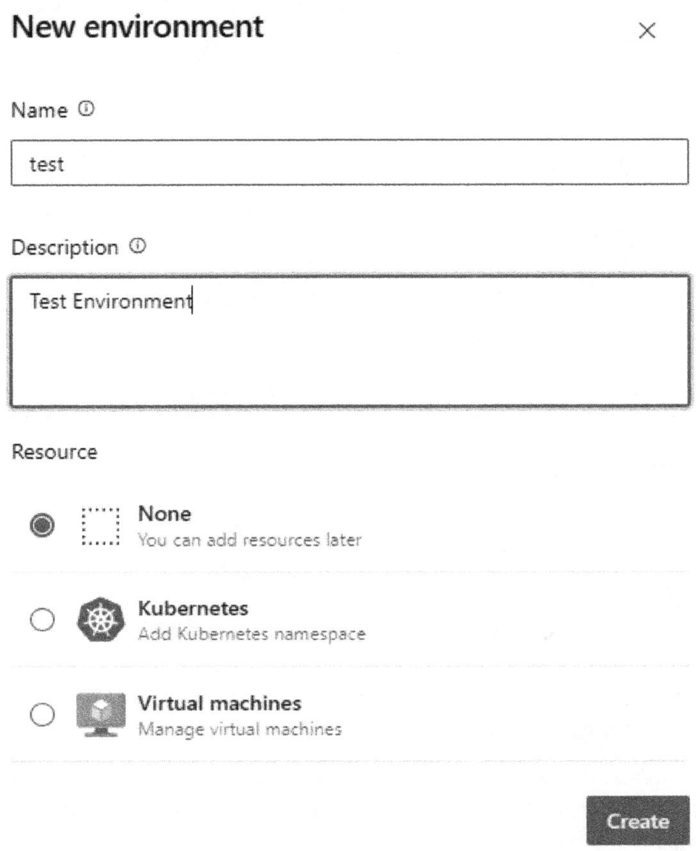

Figure 9.9 – Creating the test environment

4. Repeat *steps 2 and 3* to create another environment using `production` for **Name** and `Production Environment` for **Description**.

5. With the two environments created, click on the **production** one in the list, as shown here:

Figure 9.10 – Environments

6. In the next screen, you are going to add an approval check, to ensure you can only deploy to production once a human indicates it is possible.

For this, start by clicking on the ellipsis button in the top-right part of the screen and then on the **Approvals and checks** item, as shown here:

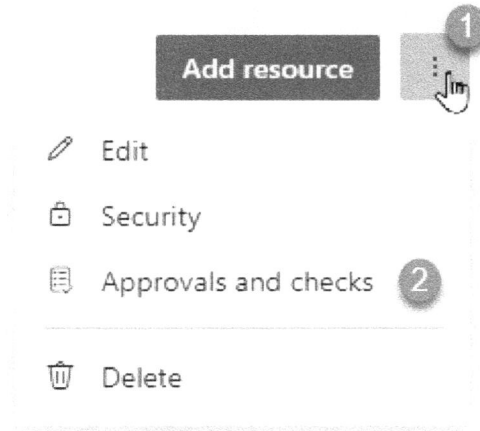

Figure 9.11 – Adding an environment approval gate

7. Since no checks have been added yet, you should see a screen like the following; selecting **Approvals** will get us to the next step to complete the check configuration:

Add your first check

Checks allow you to manage how this resource is used.

Changes made to checks are effective immediately, applicable to all existing and new pipelines.

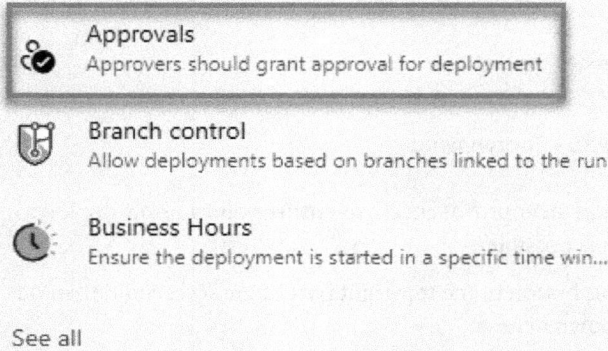

> **Approvals**
> Approvers should grant approval for deployment

> **Branch control**
> Allow deployments based on branches linked to the run

> **Business Hours**
> Ensure the deployment is started in a specific time win...

See all

Figure 9.12 – Adding an approval check

In this form, provide the required approvers (it could be yourself initially). You can optionally provide instructions, such as manual steps to verify by the approver, and change **Timeout**. If you are the approver, you must make sure the **Allow approvers to approve their own runs** option is checked under **Advanced**. Finally, hit the **Create** button and you will be ready to move on to the next steps:

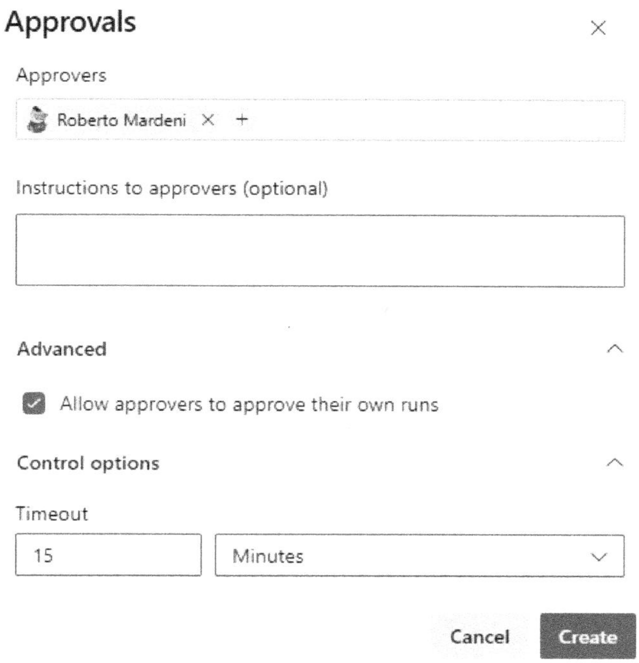

Figure 9.13 – Creating an approval check

8. Lastly, you must also add permissions for each environment to allow them to be used in the **E2E-Azure** pipeline. Like before, click on the ellipsis option in the top-right part of the screen and click the **Security** option from the menu.

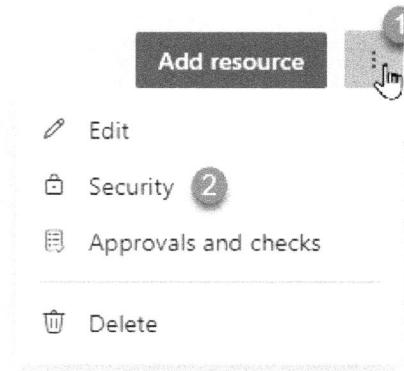

Figure 9.14 – Environment security settings

9. Then click on the + button and search for the **E2E-Azure** pipeline to add the permissions; just click on the name to add it.

Pipeline permissions

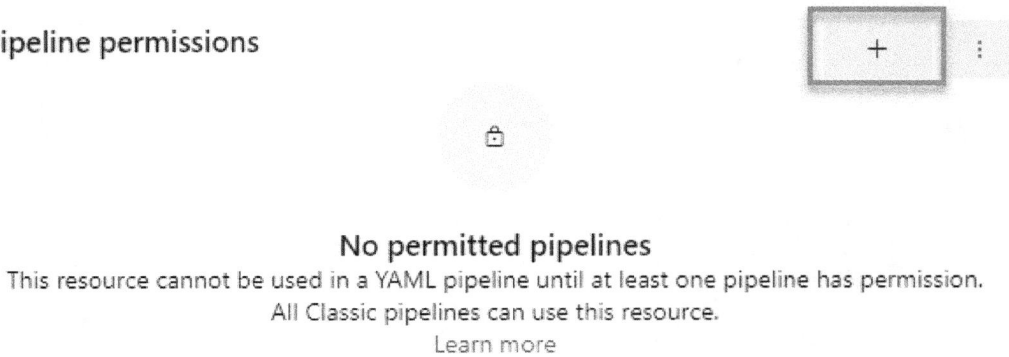

No permitted pipelines
This resource cannot be used in a YAML pipeline until at least one pipeline has permission. All Classic pipelines can use this resource.
Learn more

Figure 9.15 – Adding pipeline permissions to environment

A properly configured environment will look like the following:

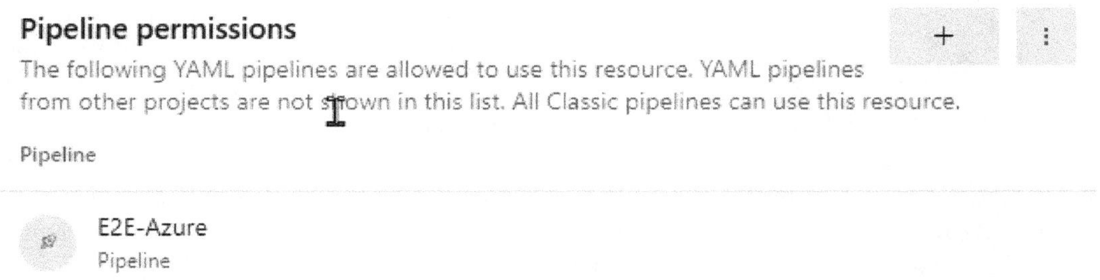

Pipeline permissions

The following YAML pipelines are allowed to use this resource. YAML pipelines from other projects are not shown in this list. All Classic pipelines can use this resource.

Pipeline

E2E-Azure
Pipeline

Figure 9.16 – Environment with pipeline permissions

Important

It is a good practice to create environments for all deployment stages; this allows us to be modular and templatize deployment steps, giving us the opportunity to add approvals or gates later if need be.

Now that you have our environments configured, let's move on to the deployment steps.

Deploying to environments

You will deploy the environment by creating a `deploy.yml` file and start by adding the steps needed for AKS deployment and the Python catalog service.

The `deploy.yml` file will start with the following content; you will be adding to it in every section hereafter:

```
parameters:
- name: envName
  type: string
  default: 'test'
- name: azureSubscription
  type: string
  default: 'azure-packt-rg'
- name: resourceGroupName
  type: string
  default: 'packt'
- name: location
  type: string
  default: 'East US'

jobs:
- deployment: deployment_${{ parameters.envName }}
  displayName: Deploy to ${{ parameters.envName }}
  environment: ${{ parameters.envName }}
  strategy:
    runOnce:
      deploy:
        steps:
```

So far, you don't have much in this file, but let's break it down:

- The `parameters` section defines all the values available for reuse within the pipeline definition. The only one being used from the main pipeline is the `envName` one, but this gives you the flexibility to change them when needed.

- The `jobs` collection includes a new job type you haven't used before called `deployment`, which allows us to implement different rollout strategies. For simplicity, here you will be using the `runOnce` strategy, but you can also use `canary` and `rolling` where appropriate. To learn more about these, go to `https://learn.microsoft.com/en-us/azure/devops/pipelines/yaml-schema/jobs-deployment-strategy`.

Now, you can proceed with the first service deployment steps.

Deploying the Python catalog service to AKS

The deployment of the Python catalog service to Azure Kubernetes Service is performed as a two-phase process:

1. Deploy the ARM template to create and configure the AKS cluster using IaC. Refer to *Chapter 8* for more information on how to do this.

2. Deploy the application using the Helm chart provided in the repository.

In our `deploy.yml` file, you will add the following six steps:

1. The `download` task `Download catalog iac` retrieves the pipeline artifact:

```
- download: current
  displayName: 'Download catalog iac'
  artifact: catalog-iac
```

2. The `AzureResourceGroupDeployment@2` task performs the AKS deployment:

```
- task: AzureResourceGroupDeployment@2
  name: catalogInfra
  displayName: 'Deploy catalog Infra'
  inputs:
    azureSubscription: ${{ parameters.azureSubscription }}
    action: 'Create Or Update Resource Group'
    resourceGroupName: ${{ parameters.resourceGroupName }}
    location: ${{ parameters.location }}
    templateLocation: 'Linked artifact'
    csmFile: '$(Pipeline.Workspace)/catalog-iac/template.json'
    overrideParameters: '-environmentName ${{ parameters.envName }}'
    deploymentMode: 'Incremental'
    deploymentName: 'catalog-$(Build.BuildNumber)'
    deploymentOutputs: 'catalogInfraOutputs'
```

Notice that you are using the `deploymentOutputs` property to set the name of a variable that will contain the outputs generated in the ARM template; this will be needed later for the frontend deployment.

3. The `PowerShell@2` task parses out the cluster name from the output parameters of the ARM template deployment and makes it available as another variable for the duration of the job:

```
- task: PowerShell@2
  displayName: 'Get AKS clusterName'
  inputs:
    targetType: 'inline'
```

```
    script: |
      $var=ConvertFrom-Json '$(catalogInfraOutputs)'
      $value=$var.clusterName.value
      Write-Host "AKS Cluster Name: $value"
      Write-Host "##vso[task.setvariable
  variable=clusterName;]$value"
```

4. The download task Download catalog helm chart retrieves the artifact:

```
- download: current
  displayName: 'Download catalog helm chart'
  artifact: catalog-helm-chart
```

5. The HelmInstaller@1 task installs Helm in the agent:

```
- task: HelmInstaller@1
  displayName: 'Install Helm'
  inputs:
    helmVersionToInstall: 3.11.3
```

6. The HelmDeploy@0 task performs an upgrade command with the install option set to true; this will guarantee that, if it's not found, it will be created:

```
- task: HelmDeploy@0
  displayName: Deploy Catalog App to AKS
  inputs:
    connectionType: 'Azure Resource Manager'
    azureSubscription: ${{ parameters.azureSubscription }}
    azureResourceGroup: ${{ parameters.resourceGroupName }}
    kubernetesCluster: $(clusterName)
    releaseName: catalog
    chartType: FilePath
    chartPath: "$(Pipeline.Workspace)/catalog-helm-chart/packt-
  store-catalog-1.0.0.tgz"
    overrideValues: 'image.tag=$(Build.BuildNumber)'
    command: upgrade
    install: true
    waitForExecution: true
```

Notice that, in this case, you are using the overrideValues property to pass in the BuildNumber value as the container tag to use, replacing the value of image.tag in the Helm chart.

It's worth noting the usage of a special output command in this section:

```
"##vso[task.setvariable variable=name;]value"
```

This is called a **logging command** in Azure Pipelines, and if a task writes this format into the console output, it will be interpreted as a command to set the value of a variable with the given `name` and `value`.

If the variable does not exist before running the command, then it will be created and made available at runtime.

To learn more about how to work with variables in Azure Pipelines, go to `https://learn.microsoft.com/en-us/azure/devops/pipelines/process/set-variables-scripts`.

Now let's move on to the Node.js cart service.

Deploying a Node.js cart service to ACA

The deployment of the cart service to ACA is a bit simpler; it just requires the use of the `AzureResourceGroupDeployment@2` task after the artifact is downloaded:

```
- download: current
  displayName: 'Download cart iac'
  artifact: cart-iac
- task: AzureResourceGroupDeployment@2
  name: cartInfra
  displayName: 'Deploy cart Infra'
  inputs:
    azureSubscription: ${{ parameters.azureSubscription }}
    action: 'Create Or Update Resource Group'
    resourceGroupName: ${{ parameters.resourceGroupName }}
    location: ${{ parameters.location }}
    templateLocation: 'Linked artifact'
    csmFile: '$(Pipeline.Workspace)/cart-iac/template.json'
    overrideParameters: '-environmentName ${{ parameters.envName }}
-containerTag "$(Build.BuildNumber)"'
    deploymentMode: 'Incremental'
    deploymentName: 'cart-$(Build.BuildNumber)'
    deploymentOutputs: 'cartInfraOutputs'
```

Notice, in this case, the use of `overrideParameters` to pass in the value of the `containerTag` parameter using `BuildNumber`.

Next, you will add the deployment of the checkout service.

Deploying a .NET checkout service to ACI

The deployment of the checkout service is very similar to the ACA deployment; the only difference is the use of the ACI service instead. See the following steps:

```
- download: current
  displayName: 'Download checkout iac'
  artifact: checkout-iac
- task: AzureResourceGroupDeployment@2
  name: checkoutInfra
  displayName: 'Deploy checkout Infra'
  inputs:
    azureSubscription: ${{ parameters.azureSubscription }}
    action: 'Create Or Update Resource Group'
    resourceGroupName: ${{ parameters.resourceGroupName }}
    location: ${{ parameters.location }}
    templateLocation: 'Linked artifact'
    csmFile: '$(Pipeline.Workspace)/checkout-iac/template.json'
    overrideParameters: '-environmentName ${{ parameters.envName }}
 -containerTag "$(Build.BuildNumber)"'
    deploymentMode: 'Incremental'
    deploymentName: 'checkout-$(Build.BuildNumber)'
    deploymentOutputs: 'checkoutInfraOutputs'
```

Just like you did in the Node.js deployment to ACA, in this section, you used `overrideParameters` to pass in the value of the `containerTag` parameter using `BuildNumber`.

Now let's move on to the last application, the frontend.

Deploying an Angular frontend app to AAS

For the frontend application, there are a few more steps necessary because of the need to gather information before being able to use the ARM template.

You will be adding the following steps to `deploy.yaml`:

1. The `download` task retrieves the frontend pipeline artifact:

    ```
    - download: current
      displayName: 'Download frontend iac'
      artifact: frontend-iac
    ```

2. The `AzureCLI@2` task `Get Catalog App IP from AKS` is a script needed to retrieve the IP assigned to the exposed entry point of the catalog application. It uses the `az` CLI, the `kubectl` CLI, and the `jq` tool in Linux to parse out the information from Kubernetes. This is very specific to how this application was deployed. This script might not be reusable, but it is meant to show the flexibility of the tools if needed:

```
- task: AzureCLI@2
  displayName: 'Get Catalog App IP from AKS'
  inputs:
    azureSubscription: 'azure-packt-rg'
    scriptType: 'bash'
    scriptLocation: 'inlineScript'
    inlineScript: |
      az aks get-credentials -g ${{ parameters.resourceGroupName
}} -n $(clusterName) --overwrite-existing
      ip=`kubectl get service catalog-packt-store-catalog -o
json | jq ".status.loadBalancer.ingress[0].ip"`
      echo "Catalog App IP: $ip"
      echo «##vso[task.setvariable variable=catalogAppIp;]$ip"
```

3. The `PowerShell@2` task is used to parse out the fully qualified domain name contained in the output variables generated by the previous steps. It is used to deploy the catalog, cart, and checkout services:

```
- task: PowerShell@2
  displayName: 'Set App URLs'
  inputs:
    targetType: 'inline'
    script: |
      # Set the catalog app url
      $value="http://" + $(catalogAppIp) + ":5050/"
      Write-Host "Catalog App Url: $value"
      Write-Host "##vso[task.setvariable
variable=catalogAppUrl;]$value"

      # Set the cart app url
      $var=ConvertFrom-Json '$(cartInfraOutputs)'
      $value=$var.containerAppFqdn.value
      $value="https://" + $var.containerAppFqdn.value + "/"
      Write-Host "Cart App Url: $value"
      Write-Host "##vso[task.setvariable
variable=cartAppUrl;]$value"

      # Set the checkout app url
      $var=ConvertFrom-Json '$(checkoutInfraOutputs)'
```

```
    $value="http://" + $var.containerFQDN.value + ":5015/"
    Write-Host "Checkout App Url: $value"
    Write-Host "##vso[task.setvariable
variable=checkoutAppUrl;]$value"
```

4. The `AzureResourceGroupDeployment@2` task deploys the Azure App Service instance and provides all the information necessary for the service to pull in the image, including the given `BuildNumber` as the tag. There are also the application URLs necessary to be stored in the service for the application to work:

```
- task: AzureResourceGroupDeployment@2
  name: frontendInfra
  displayName: 'Deploy frontend Infra'
  inputs:
    azureSubscription: ${{ parameters.azureSubscription }}
    action: 'Create Or Update Resource Group'
    resourceGroupName: ${{ parameters.resourceGroupName }}
    location: ${{ parameters.location }}
    templateLocation: 'Linked artifact'
    csmFile: '$(Pipeline.Workspace)/frontend-iac/template.json'
    overrideParameters: '-environmentName ${{ parameters.
envName }} -containerTag "$(Build.BuildNumber)" -catalogAppUrl
$(catalogAppUrl) -cartAppUrl $(cartAppUrl) -checkoutAppUrl
$(checkoutAppUrl)'
    deploymentMode: 'Incremental'
    deploymentName: 'frontend-$(Build.BuildNumber)'
    deploymentOutputs: 'frontendInfraOutputs'
```

5. The `PowerShell@2` task `Get Frontend URL` then uses another script to parse the output of the ARM template deployment to provide it both in the logs and as a variable that could ultimately be used in additional steps, such as a web request to smoke test the endpoint. Alternatively, it could be used in automated test execution:

```
- task: PowerShell@2
  displayName: 'Get Frontend URL'
  inputs:
    targetType: 'inline'
    script: |
      # Get the frontend app url
      $var=ConvertFrom-Json '$(frontendInfraOutputs)'
      $value=$var.frontendUrl.value
      Write-Host "Frontend Url: $value"
      Write-Host "##vso[task.setvariable
variable=frontendAppUrl;]$value"
```

Wow, that was a lot of deployments, but you are not done! Once the test environment is complete, you get a chance to approve the continuation of deployment to production in the next section.

Approving environment deployments

With the deployment to the test environment complete, you should be able to see the pipeline in the **Waiting** state, as follows:

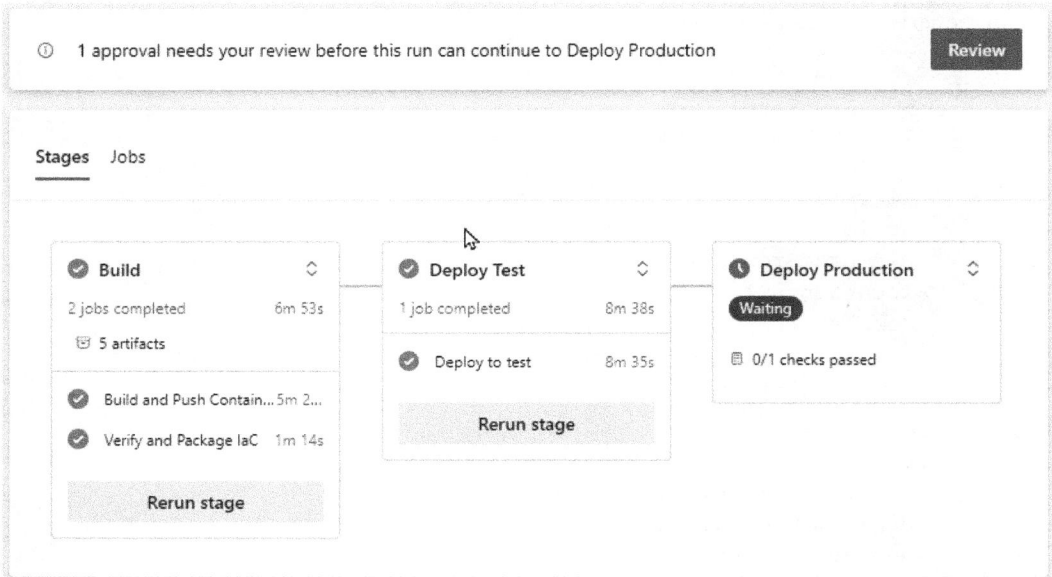

Figure 9.17 – Stage awaiting for checks

This will only look like this if you are the reviewer configured for the manual approval check. Click on the **Review** button and a new screen will pop up with **Reject** and **Approve** options and the ability to provide a comment, as shown here:

Waiting for review
Timeout in 13m   ✕

Deploy Production
Timeout in 13m

| Approved! | | Reject | **Approve** |

🕐 Approval
   Waiting for approval

🏛 production
   Environment

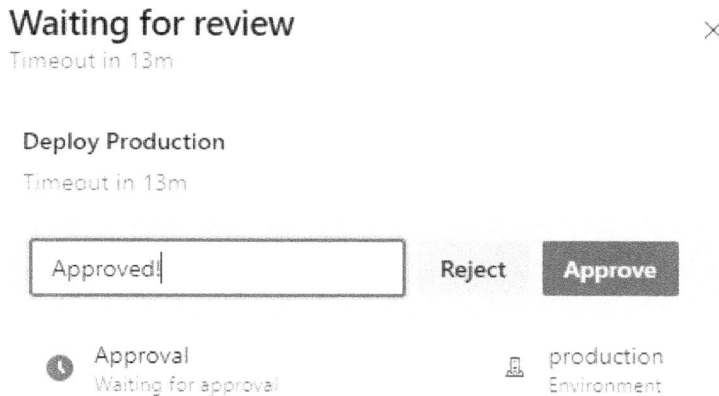

Figure 9.18 – Approving an environment check

If click the **Approve** button, the deployment will proceed. If you click the **Reject** button, the deployment will be canceled; also, if you don't do anything and the timeout runs out, the pipeline will be canceled.

Now that you have completed all the deployments, it is worth pointing out what to do if you run into issues with deployments; let's talk now about some of the typical ones.

Troubleshooting deployment issues

Creating a stable and reliable CI/CD pipeline takes time, especially when performing deployments to a cloud platform such as Azure. Let's walk through some of the typical issues you can run into.

Issues deploying IaC

You used ARM templates to deploy the infrastructure to host the services that run the applications in Azure in this chapter, which means you are relying on that infrastructure to succeed before you can deploy the applications.

There are several situations in which the `AzureResourceGroupDeployment` task can fail:

- **Internal server errors**: These can occur if the Azure region you are trying to deploy is suddenly going through capacity issues or undergoing maintenance, or even if the Azure Pipelines service is going through issues itself.

How to fix it: Usually there is no recovery from this except for attempting to run the failed pipeline again. If the region in Azure becomes unavailable, you will have to wait until it becomes available again or target a different region for deployment as part of your disaster recovery strategy.

- **Timeout**: The deployment took too long, which could have been caused by the pipeline agent or the Azure deployment. The default timeout for Microsoft-hosted agents is 60 minutes in the free tier and 360 minutes when paying for parallel jobs.

  How to fix it: You have a way to increase the timeout at the job level if required, but most likely there are other reasons why your deployment is failing. You will have to analyze the errors in the pipeline to find the root cause.

For other tips regarding deployments using ARM templates, head to `https://learn.microsoft.com/en-us/azure/azure-resource-manager/templates/best-practices`.

Issues with scripts

Scripts used in your pipelines need to be written in an *idempotent* manner, meaning that for every command or task to execute properly, the script must verify whether the operation is required and whether the result code is what is expected. This approach ensures that a script only performs the operations required to reach the desired state and, in doing so, checks every step of the way whether the operation is indeed required. Not following this approach leads to brittle scripts (scripts that are easily broken), especially when interacting with Azure resources where the current state might not match the desired state.

To address this issue, always write your scripts in an idempotent way. Follow the `if-not-then` pattern for every operation.

Issues with Helm

Here are some issues commonly observed when working with Helm:

- Helm is a very convenient tool, but you are still responsible for the proper formatting of each of the manifests and making sure that they conform with and are valid for the Kubernetes API of your cluster. In this chapter, you learned how to use **kube-linter** to validate your Helm charts, but this is only a tool, and as such it can fail to detect issues. This tool only validates against the latest stable Kubernetes API, and if your cluster is not running this version, the validation will not catch issues that will arise when performing a deployment.

  How to fix it: There are other open source tools that can validate against specific Kubernetes versions and perform different checks against your Kubernetes objects. A couple of examples are **kube-score** (`https://kube-score.com/`) and **Kubeconform** (`https://github.com/yannh/kubeconform`); put each of them to the test and evaluate which works better for your applications.

- Another issue to expect with Helm is the deployment of Kubernetes objects and the underlying consequences of this in Azure, such as deploying additional services in the case of ingress controllers. This scenario entails the creation of other Azure resources that in turn can sometimes fail.

  How to fix it: If an operation failed due to a platform timeout or retriable error, there is nothing else to do but deploy again.

With this, we've finished the chapter. Let's wrap it up.

Winding up

If you completed all the steps, you will have deployed test and production environments, so it is time to clean up! This is important because you have deployed many resources into Azure. Make sure to delete them if you do not want to keep paying for them. You can do this via the Azure portal or the following Azure CLI command:

```
az group delete -n packt -y
```

If you missed anything or got stuck and are having trouble putting the entire solution together, the complete pipeline definitions can be found in the GitHub repository mentioned in the *Technical requirements* section; look in the `ch09/azure` directory, specifically the **complete** branch.

Now, let's recap what we have learned in this chapter.

Summary

In this chapter, we took a complex solution and learned how to create CI/CD pipelines in a modular way, taking advantage of stages, environments, and templates. We also learned about adding checks throughout the stages of a pipeline. In this case, we added manual approval, but we saw that there are other controls that can be put in place to implement more complex scenarios. We learned briefly about containers and how building container-based applications with Docker Compose is easy and facilitates working with different programming languages at the same time in your pipelines; it also reduces the complexities of compiling and packaging them. We learned about semantic versioning and its applicability while learning about how build numbers can be used to tag or name artifacts from your pipelines, along with the importance of tracking artifacts. Lastly, we walked through the deployment of different services in Azure using ARM templates, learning about some of the intricacies of tying them together and the flexibility of pipelines to coordinate templates, regardless of the number of services to be deployed.

Now that we have learned how to build and deploy this complex solution to different Azure services, the next chapter will be about doing this using **Amazon Web Services** instead.

Implementing CI/CD for AWS

In this chapter, we are going to build an end-to-end solution, similar to the previous chapter, but it will target the **Amazon Web Services** (**AWS**) cloud platform, deploy the same applications, and promote them from a test environment to a production environment. This chapter showcases the flexibility of Azure Pipelines to adapt to your environment needs, no matter the destination, allowing for similar CI/CD capabilities with a different cloud provider and the ability to still be able to control the process all the way through.

We will cover the following topics:

- Explaining the solution architecture
- Building and packaging applications and IaC
- Deploying a Python catalog service to **Elastic Kubernetes Service** (**EKS**)
- Deploying a Node.js cart service to **Fargate**
- Deploying a .NET checkout service to **Elastic Container Service** (**ECS**)
- Deploying an Angular frontend app to **Lightsail**

Before we jump right in, let's take care of some technical requirements.

Technical requirements

You can find the code for this chapter at `https://github.com/PacktPublishing/Implementing-CI-CD-Using-Azure-Pipelines/tree/main/ch10`.

To complete the tasks described in this chapter, you will need the following:

- **Access to an AWS account and service connection**: It is assumed that you completed the *Access to an AWS account* and *Creating a service connection to AWS* sections in *Chapter 8*. If you skipped these, please go back and complete those steps to be able to complete this chapter.

- **The sample repository imported**: It is also assumed that you have already imported the sample repository from GitHub. If you haven't done so, check out *Chapter 9* to learn how to complete this.

> **Important note**
>
> If at any moment you get stuck while working on the pipelines, review the complete code available in the **complete** branch.

Now that we have the technical requirements covered, let's review the solution architecture.

Explaining the solution architecture

For our solution, we will use the same fictitious Packt Store from *Chapter 9*. However, in this chapter, this has been adapted to host the applications in different AWS services:

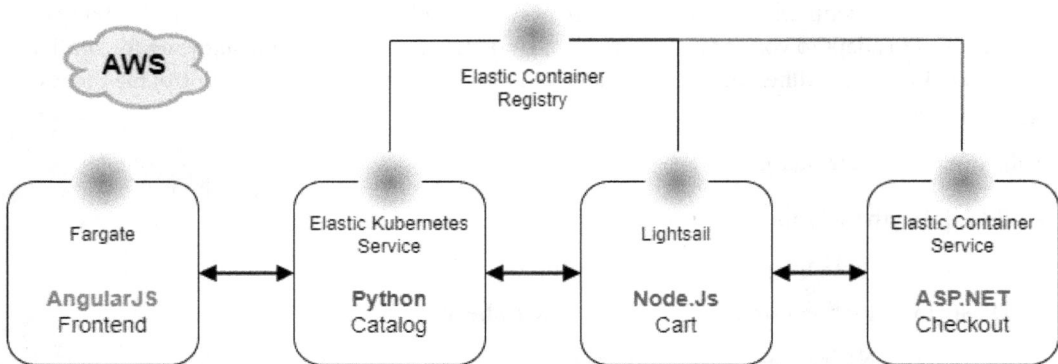

Figure 10.1 – Solution diagram

We will implement Azure Pipelines by performing the following steps for each application:

1. Build and package the application and corresponding IaC.
2. Deploy to a test environment.
3. Deploy to a production environment.
4. Automate environment deployment checks.

The following diagram depicts the CI/CD process:

Stages

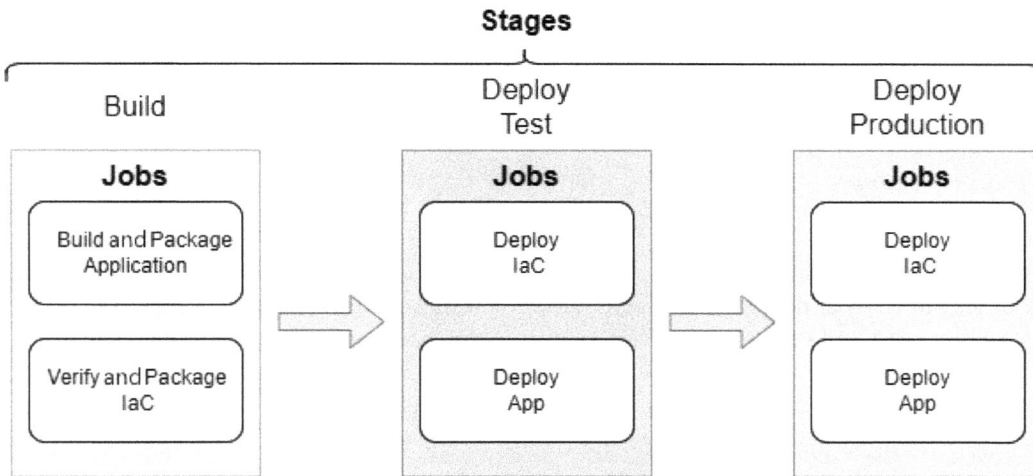

Figure 10.2 – The CI/CD process

During this chapter, we will not cover any details about the code in the applications as that is not relevant to CI/CD. Instead, we will focus on the Azure Pipelines details that are needed to make this work.

To implement the CI/CD process, we will be taking advantage of multi-stage pipelines with environments and templates, as we did in the previous chapter. Let's get started with the following pipeline definition in ch10/aws/aws-pipeline.yml. This can be found inside the Implementing-CI-CD-Using-Azure-Pipeline repository we imported:

```
# Multi-Stage pipeline
trigger:
- main
pool:
  vmImage: ubuntu-latest
stages:
- stage: build
  displayName: Build
  jobs:
  - template: build-apps.yml
  - template: build-iac.yml
- stage: deployTest
  displayName: Deploy Test
  dependsOn: build
  jobs:
  - template: deploy.yml
    parameters:
      envName: awstest
```

```
- stage: deployProduction
  displayName: Deploy Production
  dependsOn: deployTest
  jobs:
  - template: deploy.yml
    parameters:
      envName: awsproduction
```

As you can see, this pipeline definition is the same as what we used in the previous chapter, so it will work in the same way as the previous chapter. However, `build-apps.yml`, `build-iac.yml`, and `deploy.yml` will be different. Once the file is in the repository, add it as a new pipeline and rename it `E2E-AWS`. We will have to add some security configuration for everything to work at the end, such as approving the deployment to different environments, similar to what we did in *Chapter 9*.

Let's move on to the build stage.

Building and packaging applications and IaC

The applications in this solution are all container-enabled, as we saw in the previous chapter. So, in this chapter, we will go through the steps of building and pushing the container images to Amazon **Elastic Container Registry** (**ECR**) using the included `docker-compose.yml` file.

Notice that the `docker-compose.yml` file remains the same, which means that building the applications with containers allows for flexibility and ease of deployment into numerous destinations.

First, let's create the repositories that are needed in ECR.

Creating ECR repositories

By default, an ECR registry is available when you create an account in AWS, but you are responsible for creating the repositories for each of your images in it.

We can do this easily with the following AWS CLI commands:

```
aws ecr create-repository --repository-name packt-store-cart
aws ecr create-repository --repository-name packt-store-catalog
aws ecr create-repository --repository-name packt-store-checkout
aws ecr create-repository --repository-name packt-store-frontend
```

With this in place, let's move on and build the applications in the pipeline.

Creating the build apps job

Let's create the `e2e/pipelines/aws/build-apps.yml` file with the content described in this section. We have broken it into two sections for easier reading.

The first section simply defines the parameters, job header, and the only step needed. This uses the AWSShellScript@1 task for a custom script that needs to be running in the context of the AWS CLI. This must be authenticated with AWS via the service connection we created previously:

```
parameters:
- name: awsConnection
  type: string
  default: <aws-packt'
- name: location
  type: string
  default: <us-east-1>
jobs:
- job: BuildAndPushContainers
  displayName: Build and Push Containers
  steps:
  - task: AWSShellScript@1
    displayName: 'Build and Push Containers'
    inputs:
      awsCredentials: ${{ parameters.awsConnection }}
      regionName: ${{ parameters.location }}
      failOnStandardError: false
      scriptType: 'inline'
      inlineScript: |
```

With this in place, let's describe the custom script that's needed to log into ECR, build the containers, and tag and push each of them to ECR. The content of the script must be aligned properly with the preceding YAML within the inlineScript property for it to work correctly. This means that all lines must be exactly two spaces after the column where this property starts. It is only presented in this way in this chapter for ease of reading:

```
# Set variables for AWS Region and Account ID
L="${{ parameters.location }}»
ID=`aws sts get-caller-identity --query Account --output text`
# Login to ECR
aws ecr get-login-password | docker login --username AWS --password-
stdin $ID.dkr.ecr.$L.amazonaws.com
# Build images with docker-compose
docker-compose build
# Tag and push images to ECR
declare -a services=("catalog" "cart" "checkout" "frontend")
declare -a tags=("$(Build.BuildNumber)" "latest")
for s in "${services[@]}"
do
  echo «Pushing images for $s"
```

```
  for t in «${tags[@]}"
  do
    docker tag packt-store-$s:latest $ID.dkr.ecr.$L.amazonaws.com/
packt-store-$s:$t
    docker push $ID.dkr.ecr.$L.amazonaws.com/packt-store-$s:$t
  done
done
```

Let's break this code block down:

- The L and ID variables represent the AWS region and AWS account ID, both of which are needed to build the commands to log into ECR, tag images, and push them to ECR

- Logging into ECR is a two-step operation:

    A.   Retrieve a login token from the ECR service.

    B.   Use the docker login command, which allows you to use the docker and docker-compose utilities in the next steps.

- The build of the images uses the docker-compose build command, which in this context will use the existing docker-compose.yaml file in the repository

- The last step is a loop over two arrays, defining the names of services and tags to use to apply the docker tag and docker push commands

You might be wondering why we didn't use the DockerCompose@0 task like we did in *Chapter 9*. The reason for this is the types of container registries supported by this task and the authentication mechanism supported by ECR in AWS. The DockerCompose@0 task supports Azure Container Registry and the generic Docker Registry. For the latter, you could use a Docker Registry service connection, but the authorization tokens provided by ECR are short-lived, lasting only 12 hours. This would force you to update the service connection periodically.

Instead, this approach, which uses the AWSShellScript@1 task and a custom script, takes advantage of the existing AWS service connection and negotiates a new password every time it runs, storing it locally during the build phase and making it possible to have a maintenance-free setup.

Now that we have our container images available, let's work to verify and package the infrastructure as code.

Verifying and packaging IaC

We learned how to work with AWS CloudFormation templates in the previous chapter, Now, we need to validate the templates and publish them as artifacts to the pipeline.

To do this, we will create a `build-iac.yml` file in the repository and add the following six segments to it:

1. **Parameters**: These parameters will allow us to easily replace values when they're called from the main `aws-pipeline.yaml` file:

```
parameters:
- name: awsConnection
  type: string
  default: 'aws-packt'
- name: region
  type: string
  default: 'us-east-1'
```

2. **Jobs**: Here, we just add the header for the following segments, all of which are a series of steps to validate and publish the IaC artifacts that are required for the deployment of each application:

```
jobs:
- job: VerifyAndPackageIaC
  displayName: Verify and Package IaC
  steps:
```

3. **Catalog service IaC**: Add the following code:

```
  - script: docker run --rm -v $(pwd):/manifests stackrox/
kube-linter lint /manifests --config /manifests/.kube-linter.yml
    displayName: 'Lint Catalog Helm Chart'
    workingDirectory: e2e/iac/helm-charts/catalog
  - task: HelmInstaller@1
    displayName: 'Install Helm'
  - task: HelmDeploy@0
    displayName: 'Package Catalog Helm Chart'
    inputs:
      command: package
      chartPath: e2e/iac/helm-charts/catalog
      destination: $(Build.ArtifactStagingDirectory)
  - task: PublishPipelineArtifact@1
    displayName: 'Publish Catalog Helm Chart'
    inputs:
      targetPath: $(Build.ArtifactStagingDirectory)
      artifact: catalog-helm-chart
      publishLocation: 'pipeline'
```

Let's break it down:

- The script task with `displayName 'Lint Catalog Helm Chart'` validates the Helm chart
- The `HelmInstaller@1` task installs the Helm tool
- The `HelmDeploy@0` task is used to package the Helm chart
- The `PublishPipelineArtifact@1` task is then used to publish the Helm chart artifact to be used for deployment

4. **Cart service IaC**: The following code is added:

```
- task: AWSCLI@1
  displayName: 'Validate CloudFormation cart'
  inputs:
      awsCredentials: ${{parameters.awsConnection}}
      regionName: ${{parameters.region}}
      awsCommand: 'cloudformation'
      awsSubCommand: 'validate-template'
      awsArguments: '--template-body file://e2e/iac/aws/
cart/template.json'
- task: PublishPipelineArtifact@1
  displayName: 'Publish Artifacts cart'
  inputs:
    targetPath: 'e2e/iac/aws/cart'
    artifact: cart-iac
    publishLocation: 'pipeline'
```

Let's break it down:

- The `AWSCLI@1` task is used to validate the AWS CloudFormation stack template
- The `PublishPipelineArtifact@1` task is then used to publish the AWS CloudFormation stack template artifact to be used for deployment

5. **Checkout service IaC**: Add the following code:

```
- task: AWSCLI@1
  displayName: 'Validate CloudFormation checkout'
  inputs:
      awsCredentials: ${{parameters.awsConnection}}
      regionName: ${{parameters.region}}
      awsCommand: 'cloudformation'
      awsSubCommand: 'validate-template'
      awsArguments: '--template-body file://e2e/iac/aws/
checkout/template.json'
```

```
        - task: PublishPipelineArtifact@1
          displayName: 'Publish Artifacts checkout'
          inputs:
            targetPath: 'e2e/iac/aws/checkout'
            artifact: checkout-iac
            publishLocation: 'pipeline'
```

Let's break it down:

- The `AWSCLI@1` task is used to validate the AWS CloudFormation stack template

- The `PublishPipelineArtifact@1` task is then used to publish the AWS CloudFormation stack template artifact to be used for deployment

6. **Frontend application IaC**: The code to be added is as follows:

```
        - task: AWSCLI@1
          displayName: 'Validate CloudFormation frontend'
          inputs:
              awsCredentials: ${{parameters.awsConnection}}
              regionName: ${{parameters.region}}
              awsCommand: 'cloudformation'
              awsSubCommand: 'validate-template'
              awsArguments: '--template-body file://e2e/iac/aws/
    frontend/template.json'
        - task: PublishPipelineArtifact@1
          displayName: 'Publish Artifacts frontend'
          inputs:
            targetPath: 'e2e/iac/aws/frontend'
            artifact: frontend-iac
            publishLocation: 'pipeline'
```

Let's break it down:

- The `AWSCLI@1` task is used to validate the AWS CloudFormation stack template

- The `PublishPipelineArtifact@1` task is then used to publish the AWS CloudFormation stack template artifact to be used for deployment

With our IaC artifacts build phase complete, we can move on to environment deployments.

Managing environments

In this section, we will learn about how to create environments and deploy IaC and applications to them. First, let's configure our environment.

Configuring environments

As we did in *Chapter 9*, you will need to create two environments named `awstest` and `awsproduction` to complete this chapter. Once you have created these two environments, we can proceed with the deployments.

Deploying to environments

We will deploy these two environments by creating a `deploy.yml` file and start by adding the steps needed for each application. This file will start with the following content; we will be adding to it in every section hereafter:

```
parameters:
- name: envName
  type: string
  default: 'test'
- name: awsConnection
  type: string
  default: 'aws-packt'
- name: location
  type: string
  default: 'us-east-1'
- name: containerTag
  type: string
  default: '$(Build.BuildNumber)'
```

The `parameters` section defines all the values that can be reused within the pipeline definition, with `envName` being the only one used from the main pipeline, but this gives you the flexibility to change them when needed.

The `jobs` collection includes the `deployment` job type, which allows us to implement different rollout strategies:

```
jobs:
- deployment: deployment_${{ parameters.envName }}
  displayName: Deploy to ${{ parameters.envName }}
  environment: ${{ parameters.envName }}
  strategy:
    runOnce:
      deploy:
        steps:
```

For simplicity, here, we use the `runOnce` strategy, but you can also use `canary` and `rolling` where appropriate. With this in place, let's move on to the applications.

Deploying the Python catalog service to EKS

Deploying to EKS has become increasingly complex recently. To make things easier, AWS recommends using **eksctl**, an open source CLI tool created by **WeaveWorks**. This is a simple CLI tool for creating clusters on EKS that's written in the **Go** language and uses CloudFormation templates while following best practices. It handles creating or updating clusters, adding node groups, and other intermediary tasks to wait for cluster readiness that you would otherwise have to script yourself. To learn more about eksctl, go to `https://eksctl.io`.

Every section from here on is displayed aligned to the left. However, in the `deploy.yml` file, they must be aligned so that they start in the same position as the last `steps:` instruction.

Let's add the following three steps:

- The `download` task retrieves the catalog Helm chart pipeline artifact:

```
- download: current
  displayName: 'Download catalog helm chart'
  artifact: catalog-helm-chart
```

- The `HelmInstaller@1` task installs Helm in the agent:

```
- task: HelmInstaller@1
  displayName: 'Install Helm'
  inputs:
    helmVersionToInstall: 3.11.3
```

- The `AWSShellScript@1` task is used to coordinate a series of steps in a custom script while using the existing AWS service connection:

```
- task: AWSShellScript@1
  displayName: 'Deploy catalog iac and container'
  inputs:
    awsCredentials: ${{ parameters.awsConnection }}
    regionName: ${{ parameters.location }}
    arguments: '${{ parameters.envName }}-catalog ${{
parameters.location }} ${{ parameters.containerTag }}'
    disableAutoCwd: true
    workingDirectory: '$(Pipeline.Workspace)/cart-iac'
    failOnStandardError: true
    scriptType: 'inline'
    inlineScript: |
      # Confirm Parameters
      echo "SERVICE_NAME: $1"
      echo "AWS_REGION: $2"
      echo "CONTAINER_TAG: $3"
```

```
        # Install EKSCTL
        ARCH=amd64
        PLATFORM=$(uname -s)_$ARCH
        curl -sLO "https://github.com/weaveworks/eksctl/releases/
latest/download/eksctl_$PLATFORM.tar.gz"
        tar -xzf eksctl_$PLATFORM.tar.gz -C /tmp && rm
eksctl_$PLATFORM.tar.gz
        sudo mv /tmp/eksctl /usr/local/bin
        # Create Cluster
        eksctl create cluster -n $1 --version 1.27 -t t3.large -m
1 -M 2 --full-ecr-access
        # Retrieve AWS Account ID
        AWSACCOUNTID=`aws sts get-caller-identity --query Account
--output text`
        # Deploy Catalog App to EKS
        helm upgrade --install --set image.tag=${{ parameters.
containerTag }},image.repository=$AWSACCOUNTID.dkr.ecr.${{
parameters.location }}.amazonaws.com/packt-store-catalog --wait
catalog $(Pipeline.Workspace)/catalog-helm-chart/packt-store-
catalog-1.0.0.tgz
        CATALOG_HOSTNAME=`kubectl get service catalog-packt-
store-catalog -o json | jq -r ".status.loadBalancer.ingress[0].
hostname"`
        CATALOG_URL="http://$CATALOG_HOSTNAME:5050/"
        echo "Catalog URL: $CATALOG_URL"
        echo "##vso[task.setvariable
variable=CatalogUrl;]$CATALOG_URL"
```

The script does the following:

- Installs the `eksctl` tool

- Creates a simple EKS cluster

- Installs the catalog Helm chart

- Retrieves the hostname of the AWS Elastic Load Balancer that was created for the catalog service

- Sets an environment variable, `CatalogUrl`, with the properly formed URL of the catalog service to be used in the frontend deployment

Alternatively, you could store the custom script in a shell script file in the repository, use the `filePath` option for the `scriptType` property, and provide the path to the file in the `filePath` property. Refer to `https://docs.aws.amazon.com/vsts/latest/userguide/awsshell.html` for more details.

Now that we are done with the catalog service, it is time to move on to the cart service.

Deploying the Node.js cart service to Lightsail

The cart service will be deployed to the Lightsail service, which is a compute resource that's managed by AWS for running containers.

The following steps will allow you to complete the deployment:

1. The `download` task retrieves the catalog Helm chart pipeline artifact:

    ```
    - download: current
      displayName: 'Download cart iac'
      artifact: cart-iac
    ```

2. The `CloudFormationCreateOrUpdateStack@1` task creates the infrastructure required to run the service. The URL of the cart service will be automatically parsed and made available as an environment variable, in this case in the `CartUrl` variable:

    ```
    - task: CloudFormationCreateOrUpdateStack@1
      displayName: 'Create cart stack'
      inputs:
        awsCredentials: ${{ parameters.awsConnection }}
        regionName: ${{ parameters.location }}
        stackName: '${{ parameters.envName }}-cart'
        templateSource: 'file'
        templateFile: '$(Pipeline.Workspace)/cart-iac/template.json'
        onFailure: 'DELETE'
        captureStackOutputs: 'asVariables'
        captureAsSecuredVars: false
    ```

3. The `AWSShellScript@1` task performs the application deployment by doing the following:

    A. Connecting the service to the `packt-store-cart` private registry

    B. Creating a deployment with the corresponding container version

    C. Waiting for the deployment to complete by checking its state

    Let's take a look at the code that's used to do this. You will notice that it has been broken into sections for easier understanding:

    *   The following lines simply echo the parameters to the console to confirm their values while the pipeline is running:

    ```
    - task: AWSShellScript@1
      displayName: 'Deploy cart container'
      inputs:
        awsCredentials: ${{ parameters.awsConnection }}
        regionName: ${{ parameters.location }}
    ```

```
    arguments: '${{ parameters.envName }}-cart ${{ parameters.
location }} ${{ parameters.containerTag }}'
    disableAutoCwd: true
    workingDirectory: '$(Pipeline.Workspace)/cart-iac'
    failOnStandardError: true
    scriptType: 'inline'
    inlineScript: |
      # Confirm Parameters
      echo "SERVICE_NAME: $1"
      echo "AWS_REGION: $2"
      echo "CONTAINER_TAG: $3"
```

- Adding private registry access requires executing a CLI command and waiting for a property to be updated to confirm that the principal ARN has been assigned. This requires executing another CLI command to check this every 5 seconds:

```
      # Add Private Registry Access
      aws lightsail update-container-service --service-name $1
--region $2 --private-registry-access file://private-registry-
access.json
      echo "Waiting for container service to be ready..."
      principal_arn=""
      until [ "$principal_arn" != "" ]
      do
          sleep 5
          principal_arn=`aws lightsail get-container-services
--service-name $1 --region $2 --query "containerServices[0].
privateRegistryAccess.ecrImagePullerRole.principalArn" --output
text`
      done
      echo ""
      echo "Principal ARN: $principal_arn"
```

- Applying the **Elastic Container Policy (ECR)** requires deleting any existing one and then setting the new one:

```
      # Apply ECR policy
      echo "Applying ECR policy..."
      sed "s|IamRolePrincipalArn|$principal_arn|g" ecr-policy-
template.json > ecr-policy.json
      # redirect stderr to /dev/null to avoid error if policy
does not exist
      aws ecr delete-repository-policy --repository-name packt-
store-cart 2>/dev/null
      aws ecr set-repository-policy --repository-name packt-
store-cart --policy-text file://ecr-policy.json
      # Wait until container service is ready for update
```

```
        state="UPDATING"
        until [ "$state" != "UPDATING" ]
        do
            sleep 5
            state=`aws lightsail get-container-services --service-
name $1 --region $2 --query "containerServices[0].state"
--output text`
        done
        echo ""
```

- The preceding section will execute a CLI command every 5 seconds to wait for the Lightsail service to complete updates before we can proceed further. This is required because the CLI command we executed previously to assign access to the private registry takes a while to complete. This will ensure that we don't try to create a deployment while the service is still updating.

- Finally, while using a CLI command to create the deployment and waiting for it to complete in the pipeline, ensure any further steps that require the service to be running do not fail:

```
        # Create Deployment
        account_id=`aws sts get-caller-identity --query "Account"
--output text`
        sed "s|SERVICENAME|$1|g ; s|AWSACCOUNTID|$account_id|g ;
s|AWSREGION|$2|g ; s|CONTAINERTAG|$3|g" deployment-template.json
> deployment.json
        echo "Creating deployment..."
        aws lightsail create-container-service-deployment
--service-name $1 --region $2 --cli-input-json file://
deployment.json
        state="DEPLOYING"
        until [ "$state" != "DEPLOYING" ]
        do
            sleep 5
            state=`aws lightsail get-container-services --service-
name $1 --region $2 --query "containerServices[0].state"
--output text`
        done
        echo ""
        if [ "$state" == "RUNNING" ]
        then
            echo "Deployment created successfully!"
        else
            echo "Deployment failed!"
            exit 1
        fi
```

Now that we are done with the cart service, let's move on to the checkout service.

Deploying the .NET checkout service to ECS

For this, we must create a task execution IAM role. First, let's create an `ecs-tasks-trust-policy.json` file with the following content:

```
{
  "Version": "2012-10-17",
  "Statement": [
    {
      "Sid": "",
      "Effect": "Allow",
      "Principal": {
        "Service": "ecs-tasks.amazonaws.com"
      },
      "Action": "sts:AssumeRole"
    }
  ]
}
```

The following commands will create an IAM role and attach a policy that's needed to run the container image in the private registry:

```
aws iam create-role --role-name ecsTaskExecutionRole --assume-role-
policy-document file://ecs-tasks-trust-policy.json
aws iam attach-role-policy --role-name ecsTaskExecutionRole
--policy-arn arn:aws:iam::aws:policy/service-role/
AmazonECSTaskExecutionRolePolicy
```

With this complete, we can move on to the content for the `deploy.yml` file:

1. The `download` task retrieves the checkout IaC pipeline artifact:

    ```
    - download: current
      displayName: 'Download checkout iac'
      artifact: checkout-iac
    ```

2. The `CloudFormationCreateOrUpdateStack@1` task creates the infrastructure required to run the service:

    ```
    - task: CloudFormationCreateOrUpdateStack@1
      displayName: 'Create checkout stack'
      inputs:
        awsCredentials: ${{ parameters.awsConnection }}
        regionName: ${{ parameters.location }}
        stackName: '${{ parameters.envName }}-checkout'
        templateSource: 'file'
    ```

```
      templateFile: '$(Pipeline.Workspace)/checkout-iac/template.
json'
      templateParametersSource: 'inline'
      templateParameters:
'[{"ParameterKey":"ContainerTag","ParameterValue":"${{
parameters.containerTag }}"}]'
      onFailure: 'DELETE'
      captureStackOutputs: 'asVariables'
      captureAsSecuredVars: false
```

The URL of the checkout service will become available via the outputs of this deployment, which are automatically parsed and made available as an environment variable, in this case in the `CheckoutUrl` variable.

Now that we are done with the checkout service, it is time to move on to the frontend application.

Deploying the Angular frontend to Fargate

The Angular frontend application will be deployed into an ECS with a Fargate backend, a serverless offering from AWS. This deployment is much simpler because it only requires creating the CloudFormation stack, which incorporates the definition of the container to deploy within the template.

Let's add the following to the `deploy.yml` file and walk through the steps:

1. The `download` task retrieves the frontend IaC pipeline artifact:

    ```
    - download: current
      displayName: 'Download frontend iac'
      artifact: frontend-iac
    ```

2. The `CloudFormationCreateOrUpdateStack@1` task creates the infrastructure required to run the service and deploy the application:

    ```
    - task: CloudFormationCreateOrUpdateStack@1
      displayName: 'Create frontend stack'
      inputs:
        awsCredentials: ${{ parameters.awsConnection }}
        regionName: ${{ parameters.location }}
        stackName: '${{ parameters.envName }}-frontend'
        templateSource: 'file'
        templateFile: '$(Pipeline.Workspace)/frontend-iac/template.
    json'
        templateParametersSource: 'inline'
        templateParameters: '[{"ParameterKey":"ContainerTag","Pa-
    rameterValue":"${{ parameters.containerTag }}"}, {"Param-
    eterKey":"CatalogUrl","ParameterValue":"$(CatalogUrl)"},
    ```

```
{"ParameterKey":"CartUrl","ParameterValue":"$(CartUrl)"},
{"ParameterKey":"CheckoutUrl","ParameterValue":"$(Checkou-
tUrl)"}]'
    onFailure: 'DELETE'
    captureStackOutputs: 'asVariables'
    captureAsSecuredVars: false
```

Notice the different notation when providing parameters to the template in the `templateParameters` property. This is due to the way these values become available in the context of pipeline execution. When injecting values into a task, there is a distinction between pipeline parameters and variables and how they are evaluated. The `${{ parameters.name }}` notation is only processed at compile time, before runtime starts. This would be your typical usage for parameters as they should not change during runtime.

The `$(variable)` notation is processed during runtime before a task runs, which means it will be evaluated before each task is executed; any changes that are made to it through its execution will be reflected in its value. This would be your typical usage for variables. To learn more about this, read

Understand variable syntax, which is available in the official documentation at `https://learn.microsoft.com/en-us/azure/devops/pipelines/process/variables`.

With all of this in place, it's finally time to make it all work by adding the pipeline.

Adding the pipeline

Now that we have all the YAML files completed, it is time to put everything to work by adding a new pipeline. Follow these steps:

1. In the **Pipelines** section of your project, click on the **New pipeline** button, as shown in the following screenshot:

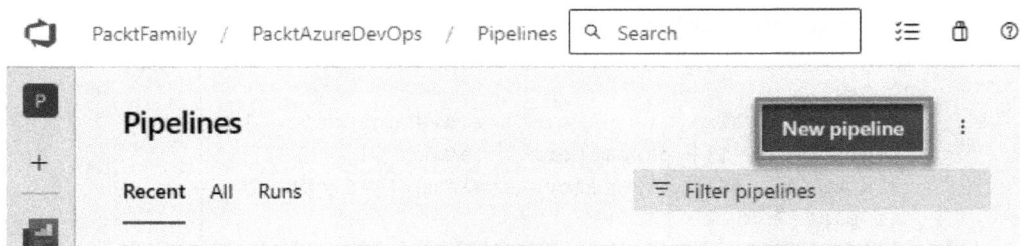

Figure 10.3 – Adding the pipeline

2. Select the **Azure Repos Git YAML** option:

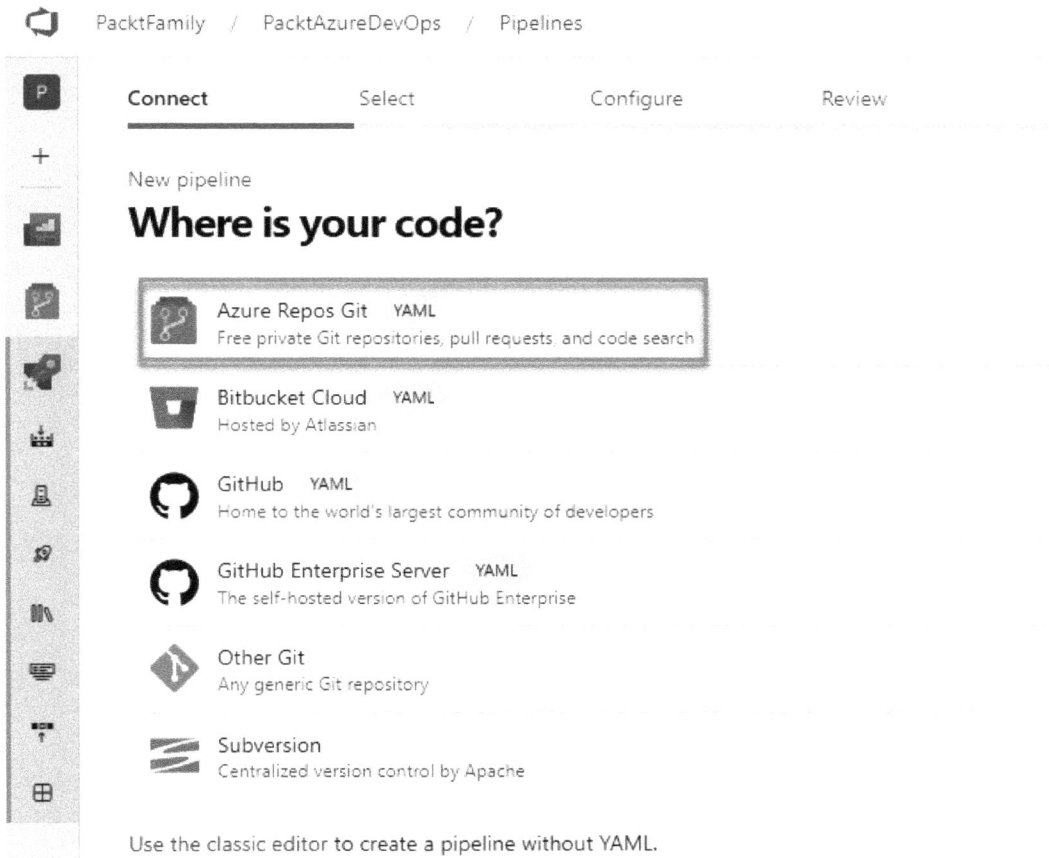

PacktFamily / PacktAzureDevOps / Pipelines

Connect   Select   Configure   Review

New pipeline

Where is your code?

Azure Repos Git   YAML
Free private Git repositories, pull requests, and code search

Bitbucket Cloud   YAML
Hosted by Atlassian

GitHub   YAML
Home to the world's largest community of developers

GitHub Enterprise Server   YAML
The self-hosted version of GitHub Enterprise

Other Git
Any generic Git repository

Subversion
Centralized version control by Apache

Use the classic editor to create a pipeline without YAML.

Figure 10.4 – Adding a pipeline from Azure Repos Git YAML

3. Select the repository where you created the pipeline:

PacktFamily / PacktAzureDevOps / Pipelines

✓ Connect   **Select**   Configure   Review

New pipeline
Select a repository

⇊ Filter by keywords   **PacktAzureDevOps** ∨   ✕

❖ IaC

❖ IaC-AWS

❖ Implementing-CI-CD-Using-Azure-Pipelines

❖ PacktAzureDevOps

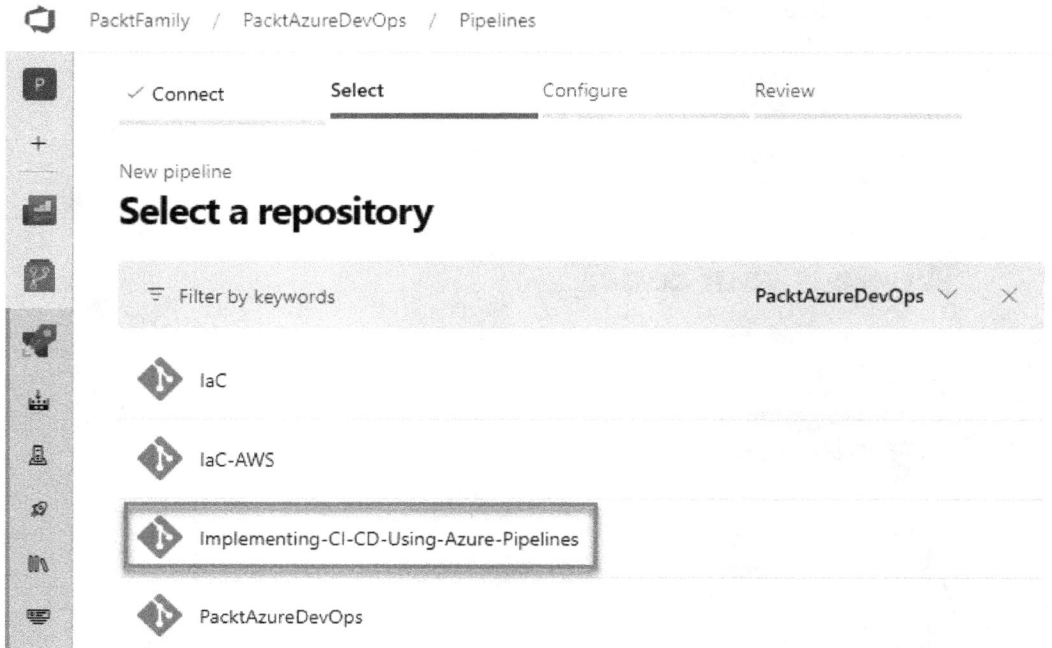

Figure 10.5 – Selecting the repository to add the YAML pipeline from

4. Then, select the **Existing Azure Pipelines YAML file** option:

PacktFamily / PacktAzureDevOps / Pipelines

✓ Connect   ✓ Select   Configure   Review

New pipeline

Configure your pipeline

Docker
docker Build a Docker image

Docker
docker Build and push an image to Azure Container Registry

Deploy to Azure Kubernetes Service
Build and push image to Azure Container Registry; Deploy to Azure Kubernetes Service

ASP.NET
Build and test ASP.NET projects.

ASP.NET Core (.NET Framework)
Build and test ASP.NET Core projects targeting the full .NET Framework.

.NET Desktop
Build and run tests for .NET Desktop or Windows classic desktop solutions.

Universal Windows Platform
Build a Universal Windows Platform project using Visual Studio.

Xamarin.Android
Build a Xamarin.Android project.

Xamarin.iOS
Build a Xamarin.iOS project.

Starter pipeline
Start with a minimal pipeline that you can customize to build and deploy your code.

Existing Azure Pipelines YAML file
Select an Azure Pipelines YAML file in any branch of the repository.

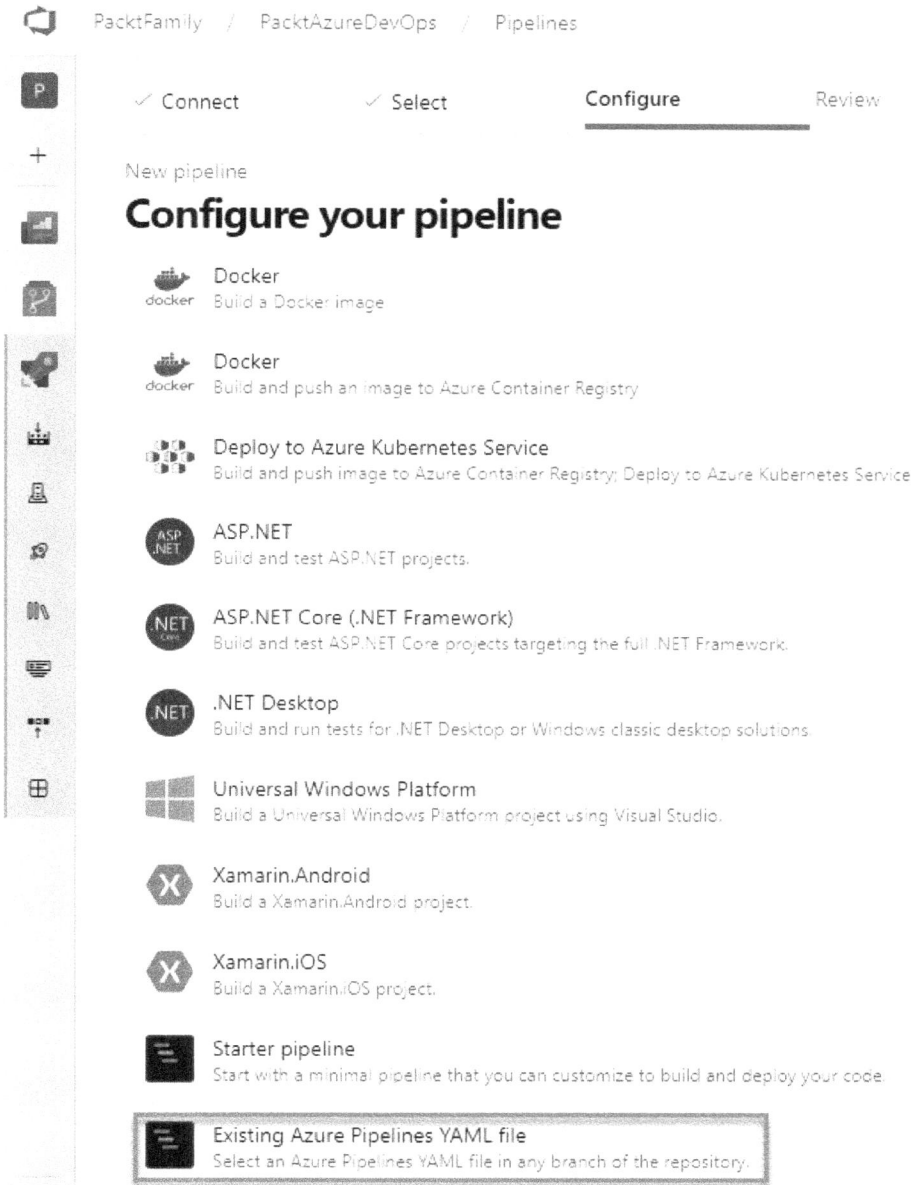

Figure 10.6 – Selecting the Existing Azure Pipelines YAML file option

5. Lastly, enter the **Path** to the pipeline file, `/e2e/pipelines/aws/aws-pipeline.yml`, and click **Continue**:

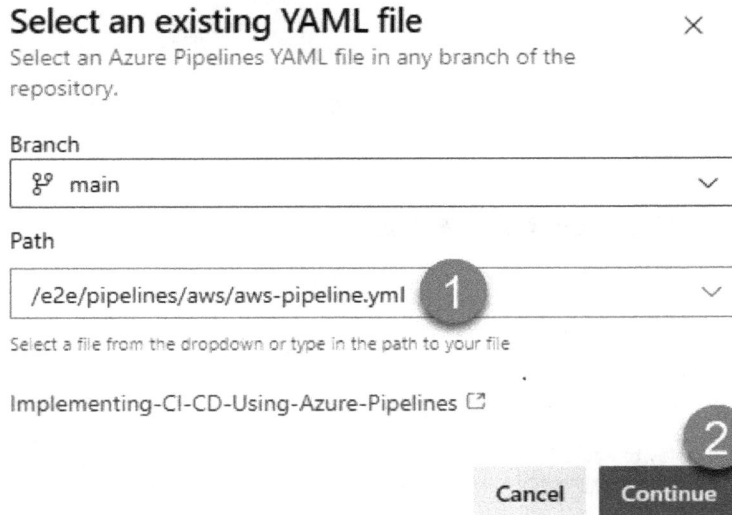

Figure 10.7 – Select an existing YAML file

With the pipeline in place, you can trigger it manually or by making changes to the repository. Now that we've got this ready, let's wrap up.

Wrapping up

If you completed all these steps, then you've deployed test and production environments, so it is time to clean up! You have deployed many resources into AWS throughout the chapter, so make sure you delete them if you do not want to keep paying for them. You can do this via the AWS console or the following AWS CLI commands:

```
eksctl delete cluster -n test-catalog
aws cloudformation delete-stack --stack-name test-cart
aws cloudformation delete-stack --stack-name test-checkout
aws cloudformation delete-stack --stack-name test-frontend
eksctl delete cluster -n production-catalog
aws cloudformation delete-stack --stack-name production-cart
aws cloudformation delete-stack --stack-name production-checkout
aws cloudformation delete-stack --stack-name production-frontend
```

If you missed anything or got stuck and are having trouble putting the entire solution together, the pipeline definitions can be found in the GitHub repository mentioned in the *Technical requirements* section, in the e2e/pipelines/aws directory in the **complete** branch.

Now, let's recap what we have learned in this chapter.

Summary

In this chapter, we learned how to deploy containerized applications to different services in the AWS cloud. At the same time, we learned how containers allow for portability across cloud providers and the ability to take advantage of multiple services within the same ecosystem.

Next, we learned how to use AWS ECR and private repositories to manage all our container images and how the process to build and push those containers, although based on the same `docker-compose` tool, must be implemented differently depending on the target platform.

We also learned about the eksctl CLI tool, which makes it easier to provision and configure EKS clusters in AWS with best practices, as well as how to use Helm charts to deploy a containerized application to a Kubernetes-based service regardless of the underlying infrastructure.

Finally, we learned how to deploy to ECS with both Fargate (serverless) and EC2 (virtual machines) infrastructure, both with a very similar and simple application deployment model.

In the next chapter, you will learn about CI/CD for **cross-mobile applications**.

11

Automating CI/CD for Cross-Mobile Applications by Using Flutter

In the previous chapter, we learned how to create a pipeline to deploy a containerized web application on AWS. This chapter will take a deep dive into creating a pipeline to automate CI/CD for the Flutter mobile application. Flutter is the most famous software development kit for mobile applications. Developers can write mobile applications using only Flutter code instead of Kotlin code for Google and Swift code for Apple, and the Azure pipeline can build and deploy Flutter code to the Google and Apple stores. They are the most widely used stores in the context of mobile applications. By the end of this chapter, you will have learned how to create a pipeline using YAML to deploy a Flutter application on Google Firebase, with the Google Play Console, and in an Apple environment.

We will cover the following topics:

- Explaining the solution architecture
- Implementing Google Firebase for Flutter
- Implementing an Apple environment for Flutter
- Implementing a Google Play Console environment for Flutter
- Navigating common challenges

Technical requirements

You will find the code for this chapter at `https://github.com/PacktPublishing/Implementing-CI-CD-Using-Azure-Pipelines/tree/main/ch11`.

To complete the tasks described in this chapter, you will need to do the following:

- Create a Firebase account and create a Firebase project by following the instructions in the official Firebase guides: `https://firebase.google.com/docs/guides`

- Set up Flutter by following the instructions provided in the official Flutter guides: `https://docs.flutter.dev/get-started/install`

- Download the Flutter code example from the GitHub repository for this book: `https://github.com/PacktPublishing/Implementing-CI-CD-Using-Azure-Pipelines`

- Create and set up your app with the Google Play Console by following the instructions in the official Google guides: `https://support.google.com/googleplay/android-developer/answer/9859152?hl=en`

- Create and set up your app in Apple Developer by following the instructions in the official Apple guides: `https://developer.apple.com/help/app-store-connect/create-an-app-record/add-a-new-app/`

Explaining the solution architecture

The following is a solution diagram that depicts how, in an Azure Pipelines workflow, to build your Flutter code and deploy it to Google Firebase, the Google Play Console, and Apple Store Connect:

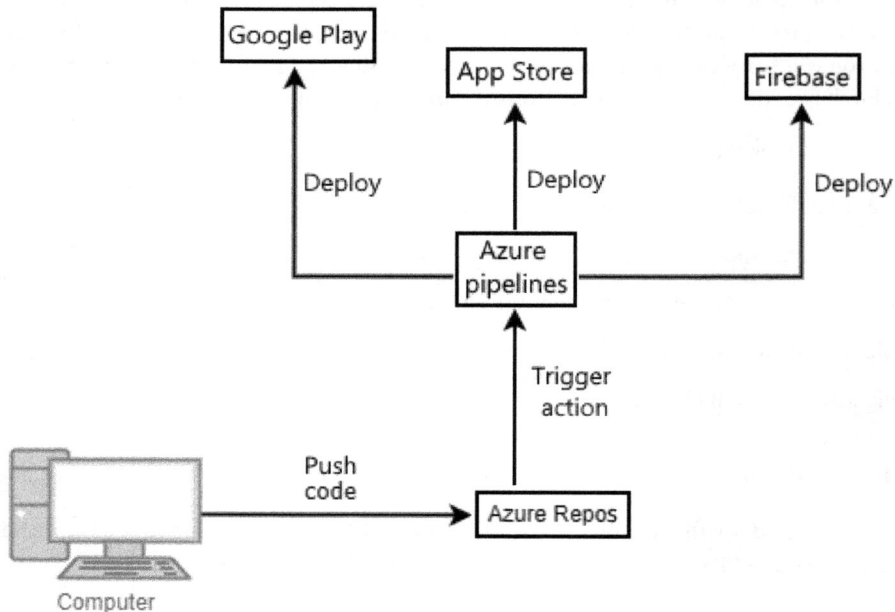

Figure 11.1 – Solution diagram

We will create three pipelines for Flutter in development and production environments, as shown in the preceding diagram, because mobile applications need to be tested on internal user or customer environments, such as Google Firebase, before deploying to production in the Google and Apple stores. The solution diagram depicts the following steps:

1. Developers develop and test mobile applications using Flutter on their machines and push Flutter code to Azure Repos.

2. After the Flutter code is uploaded to Azure Repos, Azure Repos will trigger a pipeline to build the Flutter code and deploy the Flutter applications to Google Firebase.

3. After testing the Flutter applications on Google Firebase, the developers will trigger another pipeline to build the Flutter code and deploy the Flutter applications to the Google Play Store.

4. After testing the Flutter applications on the Google Play Store, developers will trigger another pipeline to build Flutter code and deploy Flutter applications to Apple Store Connect.

Before we build our Flutter mobile application and deploy it to the Google and Apple stores, we need to generate and upload certain secure files. Let's look at these next.

Managing secure files

Before we build our Flutter mobile application and deploy it to the Google and Apple stores, we need to upload all the required files that tell the Google and Apple stores who you are, to check your developer profiles before they accept mobile applications. To do this, navigate to the **Pipelines** | **Library** | **Secure files** section and upload the following secure files to build and deploy the Flutter application:

*   You can generate an Apple certificate file based on the instructions provided at `https://developer.apple.com/help/account/create-certificates/create-developer-id-certificates` (`distribution.cer`)

*   Create a provision profile based on the instructions at `https://developer.apple.com/help/account/manage-provisioning-profiles/create-an-app-store-provisioning-profile` (`Hello_Flutter_AppStore.mobileprovision`)

*   You can download the API key file (`AuthKey_XXXXXXX.p8`) from Apple Store Connect at `https://appstoreconnect.apple.com/access/api`, as shown in the following screenshot:

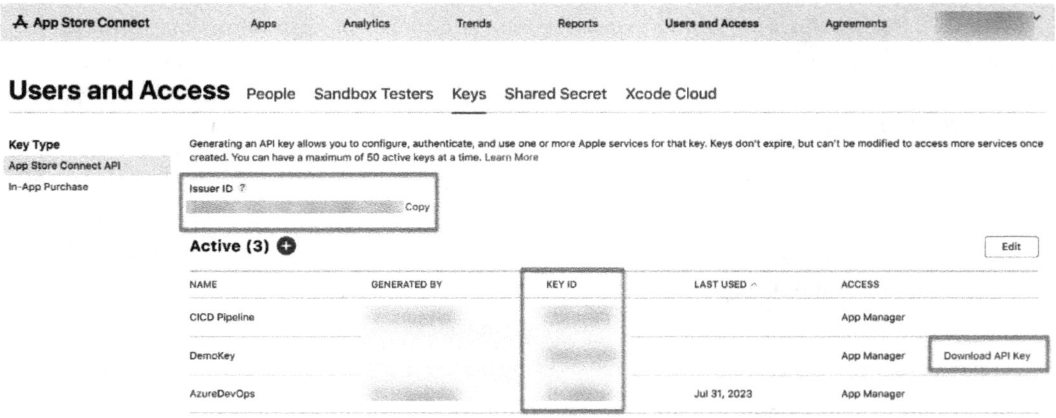

Figure 11.2 – Generate API key

Once these files are uploaded, you will see a screen similar to the following:

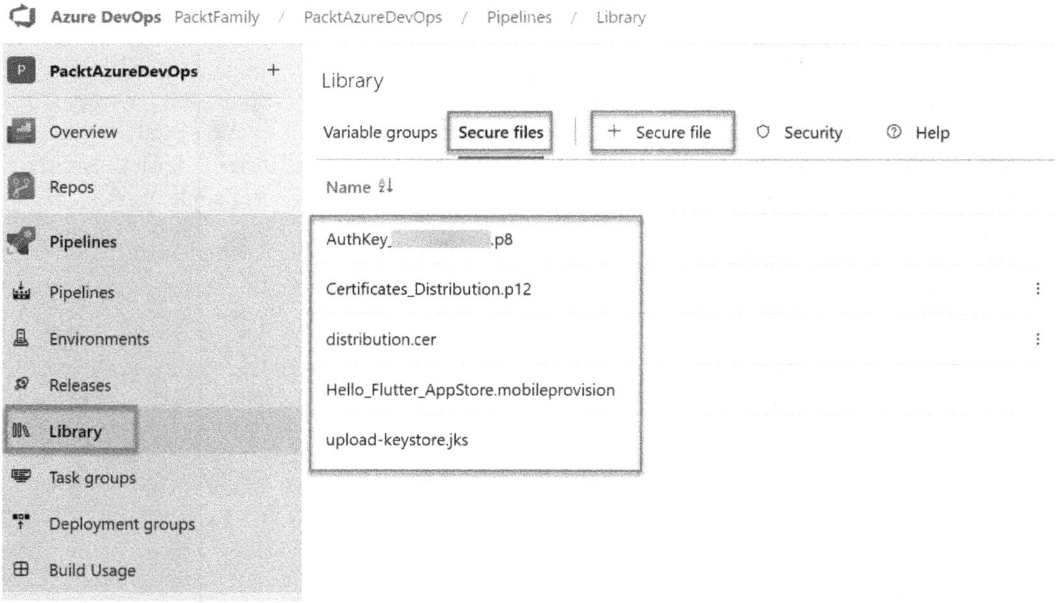

Figure 11.3 – Secure files

The following steps will show you how to prepare all the secure files shown in the preceding screenshot for Flutter application deployment:

1. **Personal Information Exchange file** (`Certificates_Distribution.p12`): For Mac, navigate to **Keychain Access** at `https://support.apple.com/en-gb/guide/keychain-access/kyca1083/mac` and the **My Certificates** tab. After that, select **Apple Distribution** (`distribution.cer`) to export it into a file with a different format, called `Certificates_Distribution.p12`.

2. **Key Store file** (`upload-keystore.jks`): Run the following command to generate a key store file:

```
keytool -genkey -v -keystore upload-keystore.jks -keyalg RSA
-keysize 2048 -validity 10000 -alias upload
```

3. Create a secret group using the **Library** option of the Azure DevOps portal:

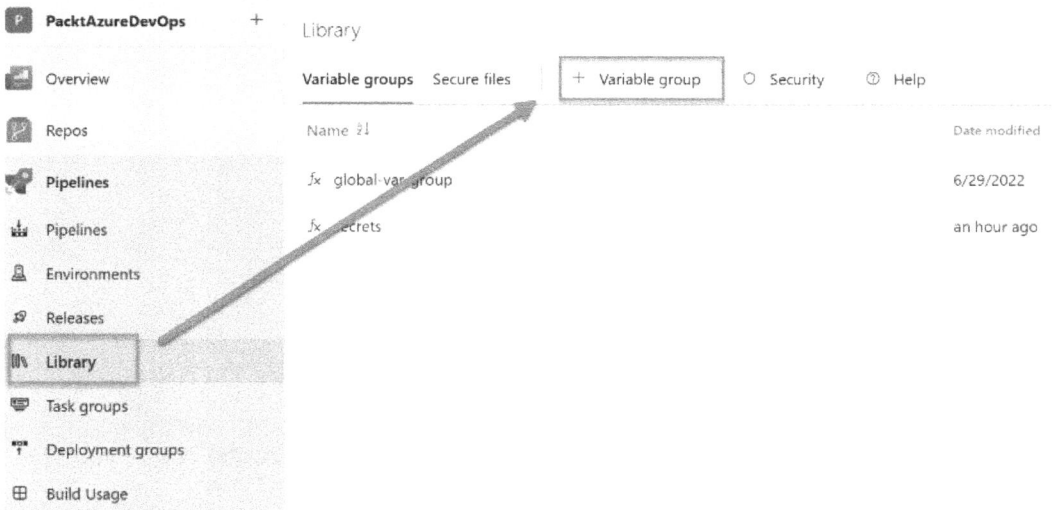

Figure 11.4 – Add a new variable group

4. Create variables for deploying Android and iOS on Google Firebase, the Google Play Console, and the App Store:

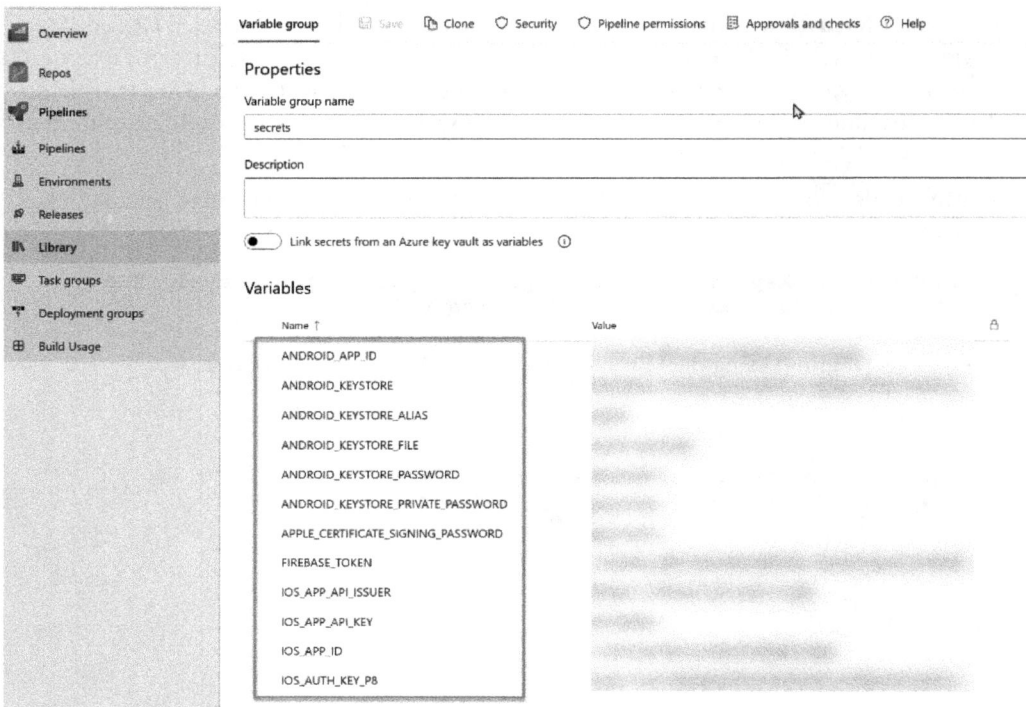

Figure 11.5 – Add all variables

Let's discuss the variables in some detail:

*   **ANDROID_APP_ID**: Navigate to the Google Firebase project page, and you will find the app ID for the Android project:

Your apps

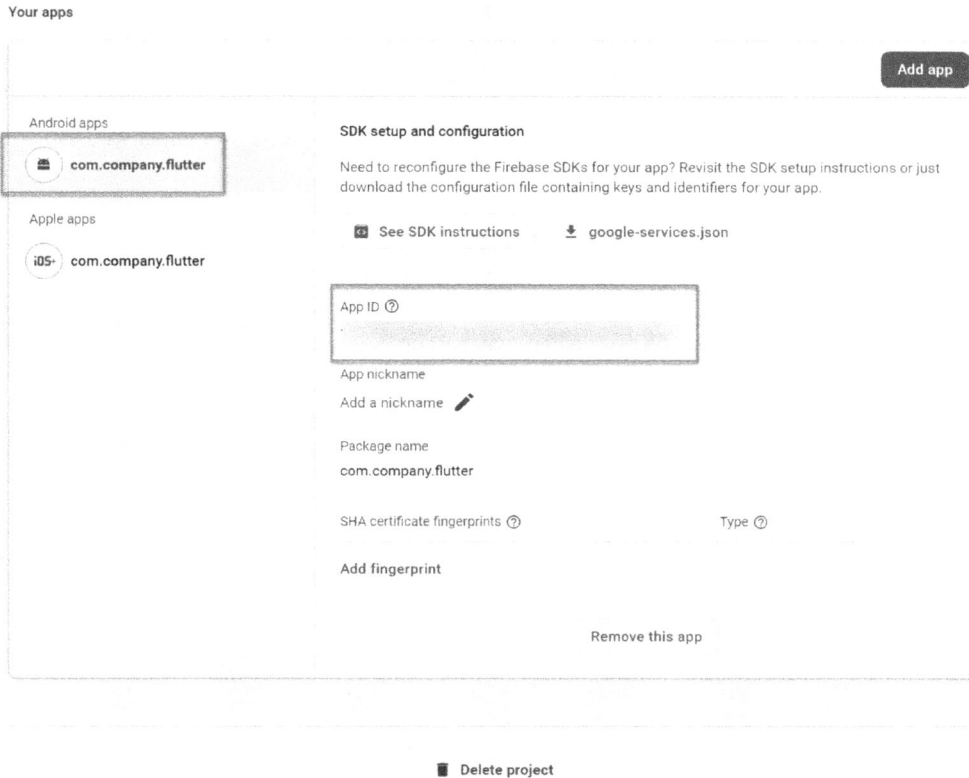

Figure 11.6 – App ID of Android project

- **ANDROID_KEYSTORE**: The following commands are used to generate it, after which you can copy the contents into the data.b64 file:

```
keytool -genkey -v -keystore upload-keystore.jks -keyalg RSA
-keysize 2048 -validity 10000 -alias upload
certutil -encode upload-keystore.jks tmp.b64 && findstr /v /c:-
tmp.b64 > data.b64
```

- **ANDROID_KEYSTORE_ALIAS**: The value should be upload.

- **ANDROID_KEYSTORE_FILE**: The value should be upload-keystore.jks. This is a **Java KeyStore** (**JKS**) file used for signing Android applications.

- **ANDROID_KEYSTORE_PASSWORD**: The password that you enter when you use the key store file generated in *step 2*.

- **ANDROID_KEYSTORE_PRIVATE_PASSWORD**: The password that you enter when you use the key store file.

- **APPLE_CERTIFICATE_SIGNING_PASSWORD**: The password when you export the P12 file.

- **FIREBASE_TOKEN**: Run the following command to generate it, after which you can copy the Firebase token on the command line result:

```
firebase token:ci
```

- **IOS_APP_API_ISSUER**, **IOS_APP_API_KEY** and **IOS_AUTH_KEY_P8**: Access the Apple Store Connect page at https://appstoreconnect.apple.com/access/users and navigate to the **Keys** section, where you can create a new API key and then download the API key file with the .p8 extension:

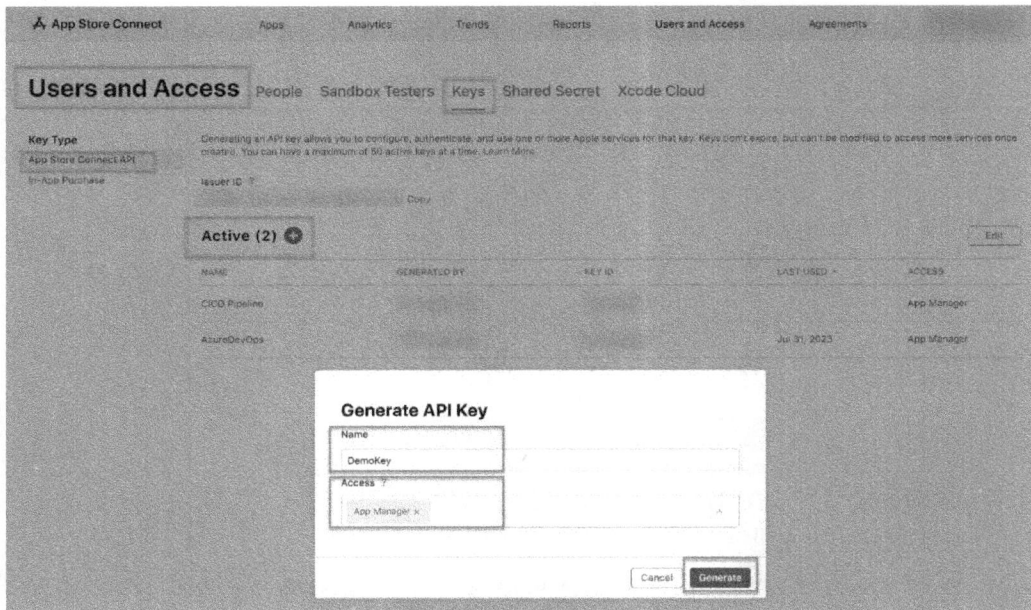

Figure 11.7 – Generate API Key

5. Next, identify the **Issuer ID**, **KEY ID**, and **API Key** file for Apple application deployment:

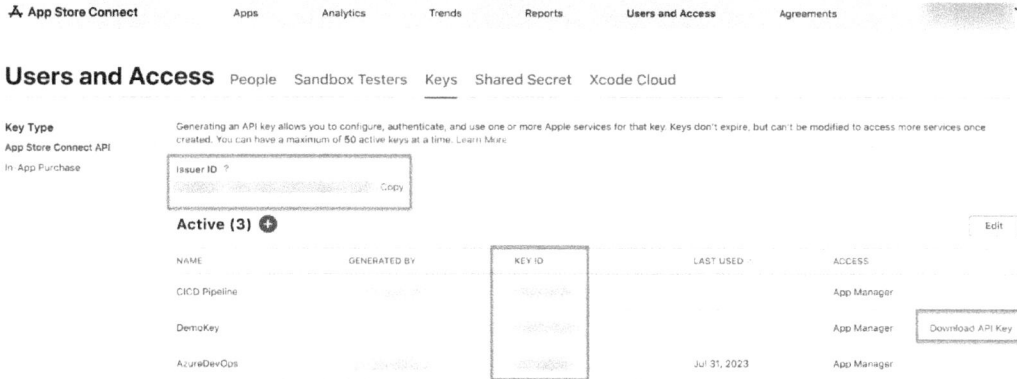

Figure 11.8 – Find Issuer ID, KEY ID, and API KEY file

6. Run the following command to convert the `.p8` file into the encode string and copy the value of the encode string into the variable called **AUTH_KEY_P8**:

```
certutil -encode AuthKey_XXXX.p8 tmp.b64 && findstr /v /c:- tmp.
b64 > data.b64
```

7. For **IOS_APP_ID**, navigate to the Google Firebase project page, and you can find **App ID** under this section of your Apple project:

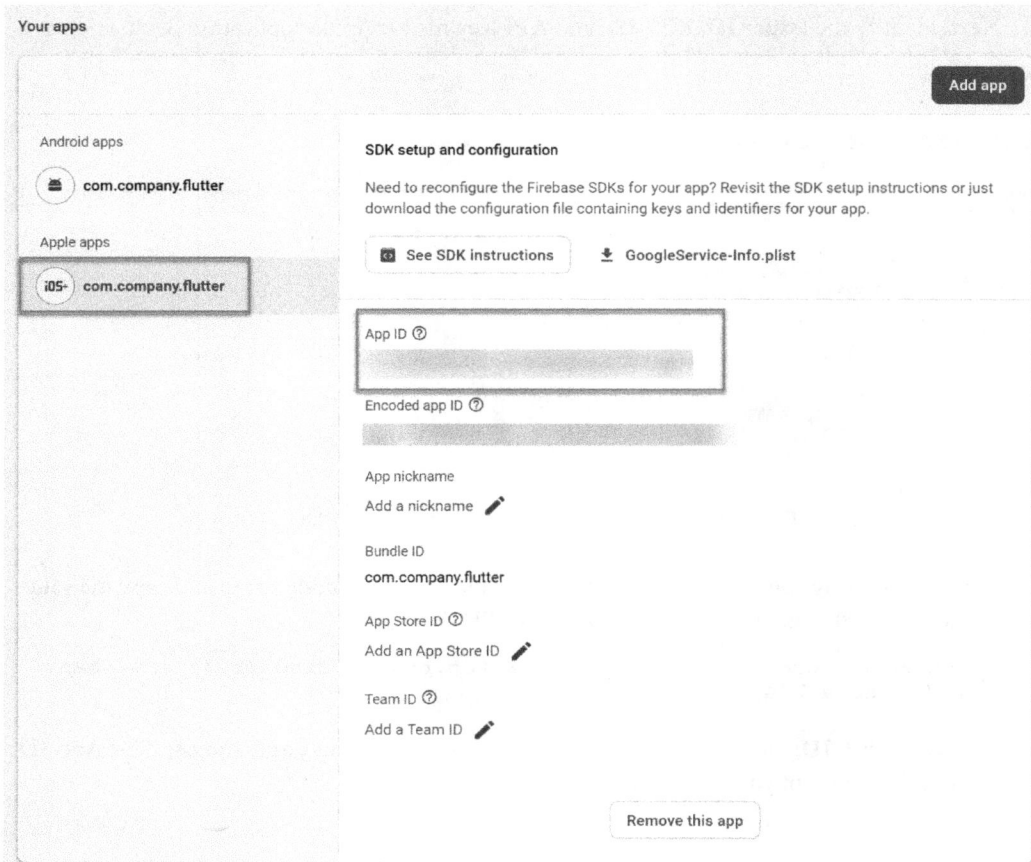

Figure 11.9 – App ID of Apple project

After preparing all the required variables and files, you can start to create a pipeline for building and deploying the Flutter application to Google Firebase, the Google Play Console, and the App Store. However, before we learn how to do this, let's discuss some key tasks that are used for deploying Flutter applications to Google Firebase, Apple environments, and Google Play.

Tasks required for Flutter applications on Google Firebase, Apple, and the Google Play Console

Here are some of the key tasks required:

- **JavaToolInstaller@0** (`https://learn.microsoft.com/en-us/azure/devops/pipelines/tasks/reference/java-tool-installer-v0?view=azure-pipelines`): This package task is for setting up the Java compiler. It is a required package task when you need to build a Flutter mobile application.

- **Hey24sheep** (`https://marketplace.visualstudio.com/items?itemName=Hey24sheep.flutter`): This package task is for setting up the Flutter compiler. It is a required package task when you need to build a Flutter mobile application.

- **CopyFiles@2** (`https://learn.microsoft.com/en-us/azure/devops/pipelines/tasks/reference/copy-files-v2?view=azure-pipelines&tabs=yaml`): This package task is for running the command to copy files.

- **PublishBuildArtifacts@1** (`https://learn.microsoft.com/en-us/azure/devops/pipelines/tasks/reference/publish-build-artifacts-v1?view=azure-pipelines`): This package task is for publishing the Flutter build files to a temporary folder named `Artifacts`.

- **Bash@3** (`https://learn.microsoft.com/en-us/azure/devops/pipelines/tasks/reference/bash-v3?view=azure-pipelines`): This package task is for running commands in a Bash shell script.

- **InstallAppleCertificate@2** (`https://learn.microsoft.com/en-us/azure/devops/pipelines/tasks/reference/install-apple-certificate-v2?view=azure-pipelines`): This package task is for installing Apple certificate files, which are used to verify Flutter applications for deploying to Apple Store Connect.

- **InstallAppleProvisioningProfile@1** (`https://learn.microsoft.com/en-us/azure/devops/pipelines/tasks/reference/install-apple-provisioning-profile-v1?view=azure-pipelines`): This package task is for installing an Apple provisioning profile file, which is used to verify a developer who is developing a Flutter application and deploying it to Apple Store Connect.

- **AppStoreRelease@1** (`https://marketplace.visualstudio.com/items?itemName=ms-vsclient.app-store#app-store-release`): This package task is for deploying Flutter applications to Apple Store Connect.

- **AndroidSigning@3** (`https://learn.microsoft.com/en-us/azure/devops/pipelines/tasks/reference/android-signing-v3?view=azure-pipelines`): This package task is for signing an Android package or APK file before deploying it to the Google Play Console.

- **GooglePlayRelease@4** (`https://marketplace.visualstudio.com/items?itemName=ms-vsclient.google-play`): This package task is for deploying Flutter applications to the Google Play Console.

- **DownloadSecureFile@1** (`https://learn.microsoft.com/en-us/azure/devops/pipelines/tasks/reference/download-secure-file-v1?view=azure-pipelines`): This package task is for downloading the secure files from the **Secure files** section of the **Variable groups** menu. In our case, it is a signing file to sign a Flutter application before deploying it on the Google Play Console.

In the next section, you will learn how to implement Google Firebase for Flutter.

Implementing Google Firebase for Flutter

To build and deploy Flutter applications on Google Firebase in both **Android** and **iOS**, you need to use the tasks discussed in the previous section. First, let's learn how we can create a pipeline for building and deploying the Flutter application to Google Firebase on Android.

Creating a pipeline for Google Firebase on Android

To create an Android pipeline, we first need to prepare an Ubuntu environment and install the Flutter compiler. After that, it will build your code into a binary file and upload it to Google Firebase App Distribution. You can follow these steps:

1. Create a pipeline file for Android called `azure-pipeline-for-firebase-android.yml` and paste the following code snippets:

    *   The first part of the YAML file is for preparing all variables that will use the entirety of this Azure pipeline. It also declares the operating system for building this file, which is Ubuntu in this scenario:

```
trigger: none
pool:
  vmImage: "ubuntu-latest"
variables:
  - group: secrets
  - name: androidReleaseDir
    value: $(build.artifactStagingDirectory)/flutter/hello_
world/build/app/outputs/flutter-apk
  - name: apkFile
    value: $(androidReleaseDir)/app-release.apk
```

    *   This task is for the pre-installation of the Java library for building your code:

```
jobs:
  - job: android_deployment
    steps:
    - task: JavaToolInstaller@0
      inputs:
        versionSpec: '11'
        jdkArchitectureOption: 'x64'
        jdkSourceOption: 'PreInstalled'
```

- This task is for the pre-installation of the Flutter compiler for building your code:

```
- task: Hey24sheep.flutter.flutter-install.FlutterInstall@0
      displayName: 'Flutter Install'
      inputs:
        version: custom
        customVersion: 3.10.6
```

- This task is for building your code and creating an APK file:

```
- task: Hey24sheep.flutter.flutter-build.FlutterBuild@0
      displayName: "Build APK"
      inputs:
        target: apk
        projectDirectory: "./flutter/hello_world"
        buildNumber: ""
```

- This task is for signing an APK file to allow Android phones to launch your application:

```
- task: AndroidSigning@3
      displayName: "Signing and aligning APK file(s) **/*.apk"
      inputs:
        apkFiles: "**/*.apk"
        apksign: true
        apksignerKeystoreFile: "upload-keystore.jks"
        apksignerKeystorePassword: "$(ANDROID_KEYSTORE_PRIVATE_
PASSWORD)"
        apksignerKeystoreAlias: "$(ANDROID_KEYSTORE_ALIAS)"
        apksignerKeyPassword: "$(ANDROID_KEYSTORE_PASSWORD)"
```

- This task is for copying a signed APK file to the artifact directory before uploading it to Azure Pipelines storage:

```
- task: CopyFiles@2
      displayName: "Copy apk to artifact directory"
      inputs:
        contents: "**/*.apk"
        targetFolder: "$(build.artifactStagingDirectory)"
```

- This task is for publishing a signed APK file to Azure Pipelines storage:

```
- task: PublishBuildArtifacts@1
      displayName: "Publish signed apk as artifact"
      inputs:
        artifactName: "drop"
```

- This task is for uploading a signed APK file from Azure Pipelines storage to Google Firebase App Distribution and letting all users download it for testing:

```
- task: Bash@3
    displayName: "Upload to firebase app distribution"
    inputs:
      targetType: "inline"
      script: |
        npm i -g firebase-tools
        ls -la $(androidReleaseDir)
        firebase appdistribution:distribute "$(apkFile)" \
          --app "$(ANDROID_APP_ID)" \
          --release-notes "Build Android From Azure Pipeline" \
          --groups "beta-testers" \
          --token "$(FIREBASE_TOKEN)"
```

2. For Android, when a pipeline build is successful, you can see the result in the **App Distribution** section of Google Firebase:

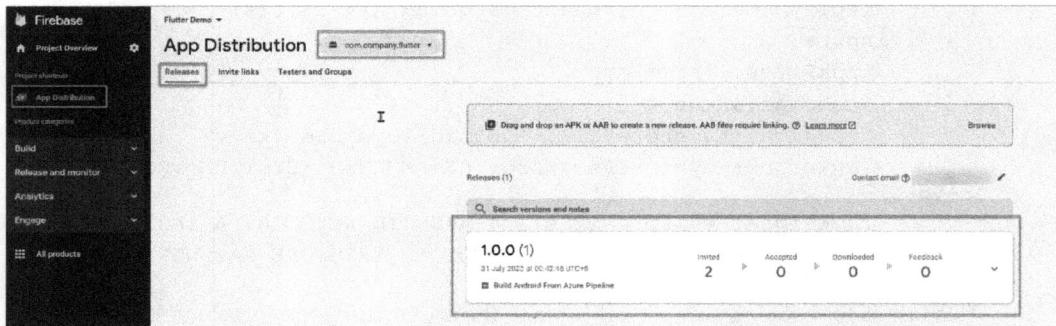

Figure 11.10 – App Distribution for the Android project

In the next section, let's learn how to do this on iOS.

Creating a pipeline for Google Firebase on iOS

To create an iOS pipeline, we first prepare a macOS environment and install the Flutter compiler. After that, it will build your code into a binary file and upload it to Google Firebase Distribution:

1. Create a pipeline file for Android called `azure-pipeline-for-firebase-ios.yml` and paste the following code snippets:

- The first part of the YAML file is for preparing all variables that will use the entirety of this

Azure pipeline. It also declares the operating system for building this file, which is the latest version of macOS:

```
trigger: none
pool:
  vmImage: "macos-latest"
variables:
  - group: secrets
  - name: iosReleaseDir
    value: $(Build.SourcesDirectory)/flutter/hello_world/build/
ios/ipa
  - name: ipaFile
    value: $(iosReleaseDir)/hello_world.ipa
  - name: rootPath
    value: $(System.DefaultWorkingDirectory)/flutter/hello_world
```

- This task is for pre-installing the Java library for building your code:

```
jobs:
  - job: ios_deployment
    steps:
    - task: JavaToolInstaller@0
      inputs:
        versionSpec: '11'
        jdkArchitectureOption: 'x64'
        jdkSourceOption: 'PreInstalled'
```

- This task is for installing Apple certificate that is required for building an iOS application:

```
- task: InstallAppleCertificate@2
    displayName: "Install Apple cert dist p12"
    inputs:
      certSecureFile: "Certificates_Distribution.p12"
      certPwd: "$(APPLE_CERTIFICATE_SIGNING_PASSWORD)"
      keychain: "temp"
```

- This task is for installing an Apple provisioning profile, which is required for building an iOS application:

```
    - task: InstallAppleProvisioningProfile@1
      displayName: "Install Apple Mobile Provisioning Profile"
      inputs:
        provisioningProfileLocation: "secureFiles"
        provProfileSecureFile: "Hello_Flutter_AppStore.
mobileprovision"
```

- This task is for installing the Flutter compiler to build your code:

```
- task: Hey24sheep.flutter.flutter-install.FlutterInstall@0
    displayName: 'Flutter Install'
    inputs:
      version: custom
      customVersion: 3.10.6
```

- This task is for building your code and creating an **iOS App Store Package** (**IPA**) file:

```
- task: Bash@3
    displayName: "Build IPA"
    inputs:
      targetType: "inline"
      script: |
        flutter build ipa --export-options-plist=$(rootPath)/
ios/Runner/ExportOptions.plist
      workingDirectory: $(rootPath)
```

- This task is for uploading an IPA file to Google Firebase App Distribution:

```
- task: Bash@3
    displayName: "Upload to firebase app distribution"
    inputs:
      targetType: "inline"
      script: |
        npm i -g firebase-tools
        ls -la $(iosReleaseDir)
        firebase appdistribution:distribute "$(ipaFile)" \
          --app "$(IOS_APP_ID)" \
          --release-notes "Build iOS From Azure Pipeline" \
          --groups "beta-testers" \
          --token "$(FIREBASE_TOKEN)"
```

2. Similar to Android, when a pipeline build is successful, you can see the result in the **App Distribution** section of Google Firebase:

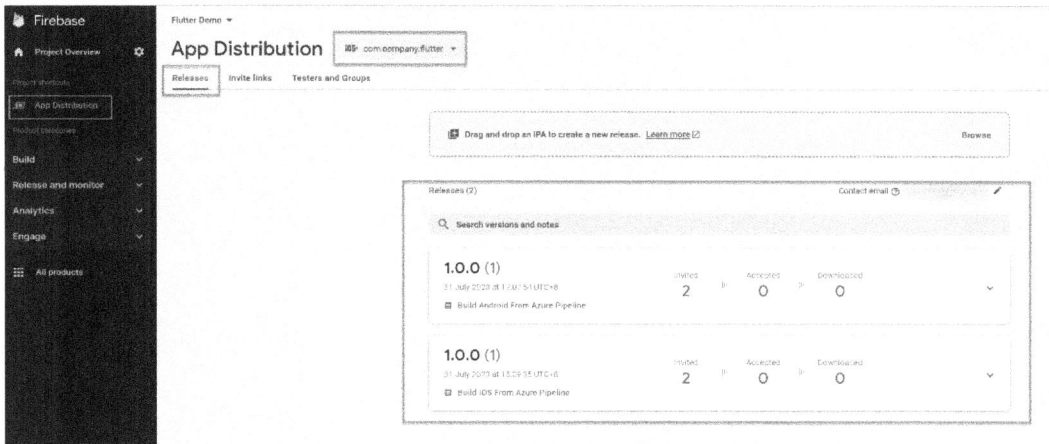

Figure 11.11 – App Distribution for the Apple project

This section discussed how to build and deploy the Flutter application on Google Firebase for testing before deploying it to the Google Play Console or App Store. Next, let's discuss how we can create and deploy Flutter applications on Apple Store Connect.

Implementing an Apple environment for Flutter

To build and deploy Flutter applications on Apple Store Connect, you need to use the various tasks in an Azure pipeline, as discussed earlier in this chapter. To reduce the time needed for mobile application deployment process, it is necessary to construct the Apple Store deployment pipeline. To do this, you can follow these steps:

1. Create a pipeline file called `azure-pipeline-for-apple-store.yml` and paste the following code snippets:

   - The first part of the YAML file is for preparing all variables that will use the entirety of this Azure pipeline. It also declares the operating system for building this file, which is the latest version of macOS:

```
trigger: none
pool:
  vmImage: "macos-latest"
variables:
  - group: secrets
  - name: iosReleaseDir
    value: $(Build.SourcesDirectory)/flutter/hello_world/build/
ios/ipa
  - name: ipaFile
```

```
      value: $(iosReleaseDir)/hello_world.ipa
   - name: rootPath
      value: $(System.DefaultWorkingDirectory)/flutter/hello_world
```

- This task is for the pre-installation of the Java library for building your code:

```
jobs:
  - job: ios_to_apple_store
    steps:
    - task: JavaToolInstaller@0
      inputs:
        versionSpec: '11'
        jdkArchitectureOption: 'x64'
        jdkSourceOption: 'PreInstalled'
```

- This task is for installing the Apple certificate that is required for building an iOS application:

```
    - task: InstallAppleCertificate@2
      displayName: "Install Apple cert dist p12"
      inputs:
        certSecureFile: "Certificates_Distribution.p12"
        certPwd: "$(APPLE_CERTIFICATE_SIGNING_PASSWORD)"
        keychain: "temp"
```

- This task is for installing an Apple provisioning profile, which is required for building an iOS application:

```
    - task: InstallAppleProvisioningProfile@1
      displayName: "Install Apple Mobile Provisioning Profile"
      inputs:
        provisioningProfileLocation: "secureFiles"
        provProfileSecureFile: "Hello_Flutter_AppStore.
mobileprovision"
```

- This task is for installing the Flutter compiler to build your code:

```
    - task: Hey24sheep.flutter.flutter-install.FlutterInstall@0
      displayName: 'Flutter Install'
      inputs:
        version: custom
        customVersion: 3.10.6
```

- This task is for building your code and creating an IPA file:

```
  - task: Bash@3
      displayName: "Build IPA"
      inputs:
```

```
        targetType: "inline"
        script: |
            flutter build ipa --export-options-plist=$(rootPath)/
    ios/Runner/ExportOptions.plist
        workingDirectory: $(rootPath)
```

- This task is for uploading a signed IPA file to Apple Store Connect:

```
- task: AppStoreRelease@1
      displayName: "Upload to App Store Connect"
      inputs:
        authType: 'ApiKey'
        apiKeyId: '$(IOS_APP_API_KEY)'
        apiKeyIssuerId: '$(IOS_APP_API_ISSUER)'
        apiToken: '$(IOS_AUTH_KEY_P8)'
        releaseTrack: 'TestFlight'
        appIdentifier: 'com.company.flutter'
        appType: 'iOS'
        ipaPath: $(ipaFile)
        shouldSkipWaitingForProcessing: true
        shouldSkipSubmission: true
        appSpecificId: '$(IOS_APP_ID)'
```

2. When a pipeline build is successful, you can see the result in the **App Store Connect | TestFlight** tab:

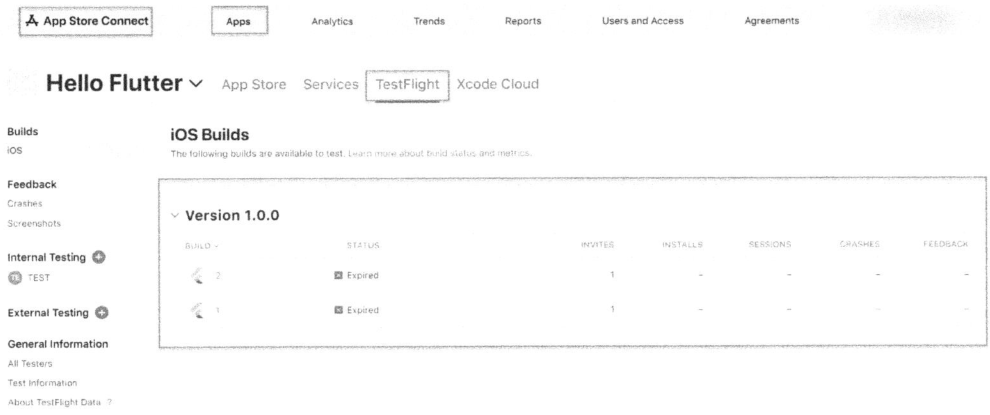

Figure 11.12 – TestFlight on App Store Connect

In this section, you learned how to create a pipeline to build and deploy Flutter applications on Apple Store Connect. In the next section, you will learn how to build and deploy Flutter applications on the Google Play Console.

Implementing a Google Play Console for Flutter

You can create an Azure pipeline when you need to upload your application to the Google Play Store, which is a marketplace for Android applications that allows all Android users to download applications. To do this, follow these steps:

1. Create a pipeline file called `azure-pipeline-for-google-play-store.yml` and paste the following code snippets:

    *   The first part of the YAML file is for preparing all variables that will use the entirety of this Azure pipeline. It also declares the operating system for building this file, which is Ubuntu:

    ```
    trigger: none
    pool:
      vmImage: "ubuntu-latest"
    variables:
      - group: secrets
      - name: androidReleaseDir
        value: $(build.artifactStagingDirectory)/flutter/hello_
    world/build/app/outputs/bundle/release
      - name: aabFile
        value: $(androidReleaseDir)/app-release.aab
    ```

    *   This task is for downloading the keystore file, which is needed to sign an **Android App Bundle** (**AAB**) file before uploading it to the Google Play Console:

    ```
    jobs:
      - job: android_to_google_play_store
        steps:
        - task: DownloadSecureFile@1
          displayName: "Download keystore file"
          name: "KeyStoreFile"
          inputs:
            secureFile: "upload-keystore.jks"
    ```

    *   This task is for pre-installing the Java library for building your code:

    ```
        - task: JavaToolInstaller@0
          inputs:
            versionSpec: '11'
            jdkArchitectureOption: 'x64'
            jdkSourceOption: 'PreInstalled'
    ```

- This task is for installing the Flutter compiler to build your code:

```
- task: Hey24sheep.flutter.flutter-install.FlutterInstall@0
  displayName: 'Flutter Install'
  inputs:
    version: custom
    customVersion: 3.10.6
```

- This task is for building Flutter code in an AAB file:

```
- task: Hey24sheep.flutter.flutter-build.FlutterBuild@0
  displayName: "Build AAB"
  inputs:
    target: aab
    projectDirectory: "./flutter/hello_world"
    buildNumber: ""
```

- This task is for copying an AAB file from the source folder to an artifact folder:

```
- task: CopyFiles@2
  displayName: "Copy aab to artifact directory"
  inputs:
    contents: "**/*.aab"
    targetFolder: "$(build.artifactStagingDirectory)"
```

- This task is for moving a signed AAB file to Azure Pipelines storage:

```
- task: PublishBuildArtifacts@1
  displayName: "Publish signed AAB as artifact"
  inputs:
    artifactName: "drop"
```

- This task is for uploading an AAB file to the Google Play Store:

```
- task: GooglePlayRelease@4
  displayName: "Upload to Google Play Store"
  inputs:
    serviceConnection: 'GooglePlayConsole'
    applicationId: 'com.company.flutter'
    action: 'SingleBundle'
    bundleFile: '$(aabFile)'
    track: 'internal'
    isDraftRelease: true
```

2.   When a pipeline build is successful, you can see the result in the Google Play Console:

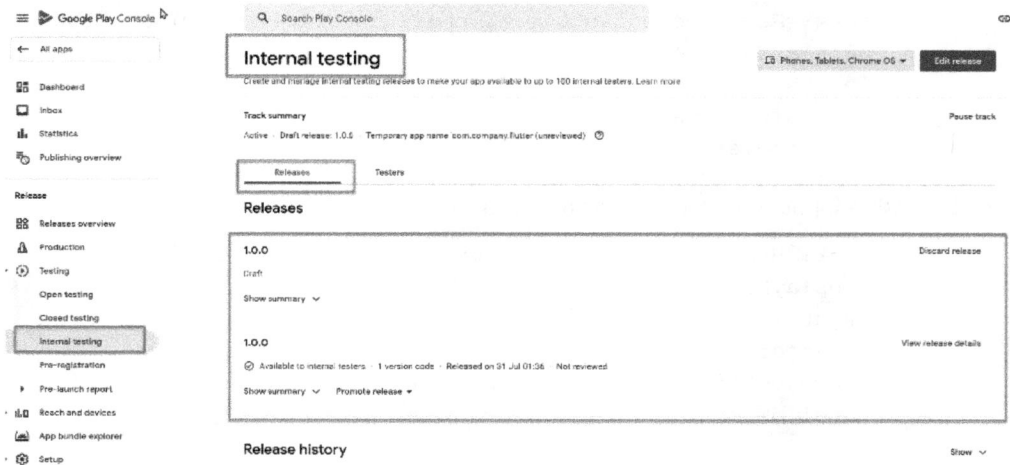

Figure 11.13 – Internal testing on the Google Play Console

In this section, you learned how to create a pipeline to build and deploy Flutter applications using the internal testing of the Google Play Console. It will help you to focus on application development after you finish setting up the pipelines.

Navigating common challenges

During Flutter development, you may face certain common issues that can be prevented if certain best practices are followed. Let's discuss a couple of them:

- **Managing dependencies**: In Flutter development, two important files are used for managing project dependencies and ensuring consistency across different development environments:

    - `pubspec.yaml`: This contains the name of the dependencies that you use for Flutter development, such as database dependencies.

    - `pubspec.lock`: This contains the name and version of dependencies that you use for Flutter development.

    Ensure that you commit `pubspec.lock` to Azure Repos because if you miss this step, you will get an error. This is because you will get a new `pubspec.lock` file, which will contain a different version of the dependencies that you used when you developed the Flutter application on your computer.

- **Handling platform-specific code**: Flutter allows you to write platform-specific code but managing this in a CI/CD pipeline can be tricky, especially when your application has custom platform-specific modules or dependencies.

  Use separate tasks in Azure Pipelines to handle platform-specific builds. For example, you might have different stages or jobs for iOS and Android builds. Ensure that the necessary SDKs and tools for each platform are installed and configured in your Azure Pipelines environment.

Now that you're well-equipped for Flutter development, let's wrap this chapter up.

Summary

This chapter taught you how to create build pipelines for Flutter applications. It covered pipelines for Google Firebase – both Android and Apple. It also covered pipelines for Apple Store Connect and Google Play Console deployment. This will help to reduce the time required by developers to create manual commands when they need to build and deploy Flutter applications. This will also help developers focus on developing mobile applications, because you create a pipeline only once and let an Azure pipeline build your code and deploy your application to Google Firebase, Apple Store Connect, and the Google Play Store automatically. This will increase developer productivity.

In the last chapter, we will learn about some common pitfalls to avoid when using Azure Pipelines and discuss the potential future applications and trends of this technology.

12

Navigating Common Pitfalls and Future Trends in Azure Pipelines

In the previous chapter, we learned how to create a pipeline for mobile applications. This chapter will teach you about the common mistakes you should avoid and, more importantly, how to effectively address and circumvent them at a high level.

We will also discuss best practices, offering insights into optimizing your workflow for maximum efficiency and reliability. This will help ensure that your pipeline operates at its full potential.

Finally, we will explore what lies ahead in the future of Azure Pipelines. By the end of this chapter, you will know how to avoid the common pitfalls of using Azure Pipelines, optimize your workflow with best practices, and adapt to the ever-evolving future of this essential tool.

In this chapter, we will cover the following topics:

- Common pitfalls
- Best practices
- Future trends

Common pitfalls

Understanding and proactively addressing certain common mistakes and issues in Azure Pipelines is essential. You can save valuable time and resources by learning how to recognize and resolve them promptly.

In the context of real-world projects, timely resolution is essential to prevent delays and costly setbacks, allowing you to maintain project momentum and focus your efforts on other critical aspects.

Let's discuss the details of these common errors:

- Sometimes, you may start the run of a pipeline and see it is queued. This could be because there is no available agent on Microsoft-hosted or self-hosted agents, insufficient concurrency, or unmet demand for agents. You can see the error immediately after you see the status of the job in your pipeline, as shown in the following screenshot:

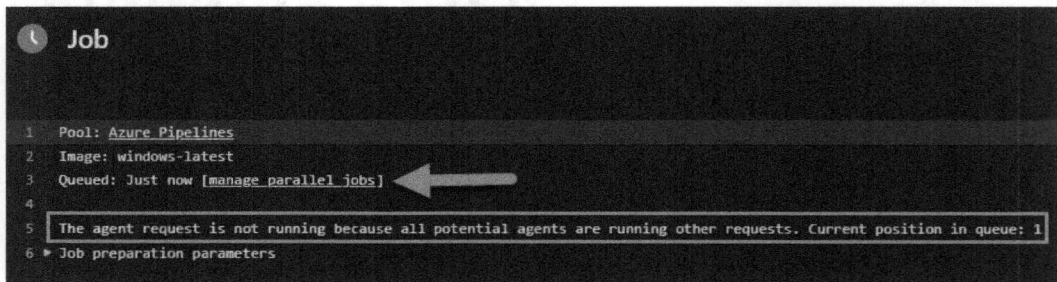

Figure 12.1 – A job queued due to unavailability of agents

This might be an indication that additional parallel jobs must be purchased, ensuring these jobs are not queued if you want these pipelines to run as quickly as possible.

- A pipeline may not run because you entered the incorrect branch for filtering in YAML files or disabled triggers. You need to go back to the YAML files and review the values under the **branch** section to ensure they are correct.

- The pipelines can't run due to the access permission for variable groups or secret files. To address this, navigate to **Library | secrets | Variable group | Pipeline permissions**:

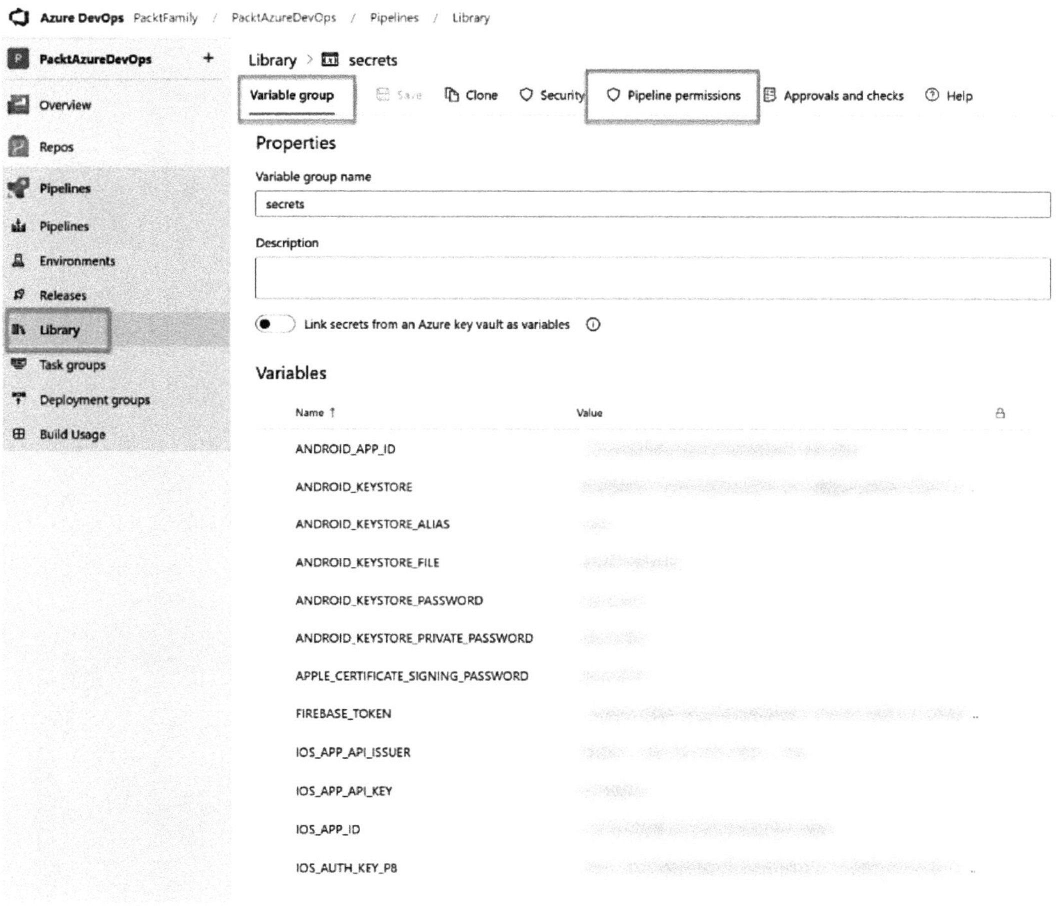

Figure 12.2 – Go to Pipeline permissions

Then, you can do either of the following:

- Open all access for all pipelines. This option will allow all pipelines access to secrets in the variable group. If all pipelines need to use the same secrets, then you should select this option:

secrets   ✕

Pipeline permissions   +   ⋮

The following YAML pipelines are allowed to use this resource. YAML pipelines f
are not shown in this list. All Classic pipelines can use this resource.   🔒 Open access

Pipeline

PacktAzureDevOps
Pipeline

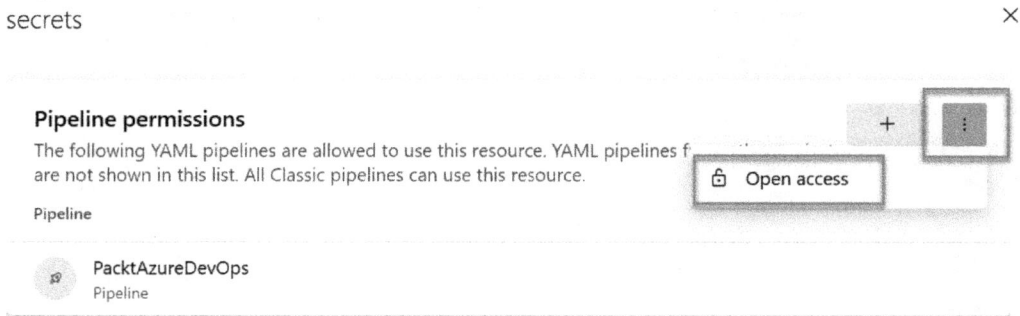

Figure 12.3 – Granting open access to variable group

- Limit access to the selected pipeline (recommended):

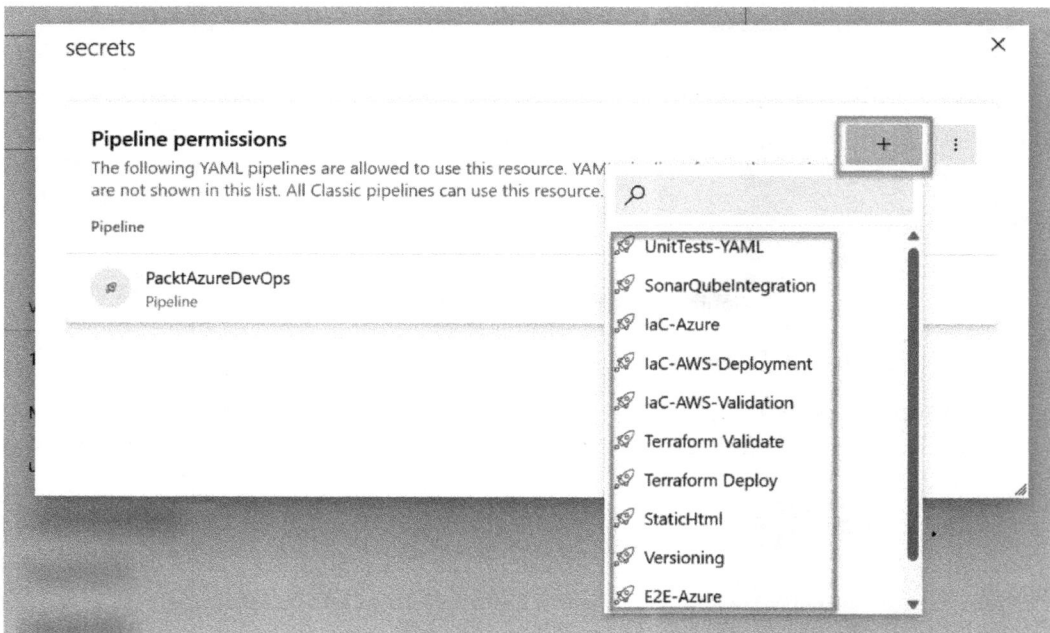

secrets   ✕

Pipeline permissions   +   ⋮

The following YAML pipelines are allowed to use this resource. YAM
are not shown in this list. All Classic pipelines can use this resource.   🔍

Pipeline   UnitTests-YAML

PacktAzureDevOps   SonarQubeIntegration
Pipeline   IaC-Azure
                                                      IaC-AWS-Deployment
                                                      IaC-AWS-Validation
                                                      Terraform Validate
                                                      Terraform Deploy
                                                      StaticHtml
                                                      Versioning
                                                      E2E-Azure

Figure 12.4 – Restricting permission to variable group to selected pipelines

This option will limit access to only the required pipelines that can access secrets because, otherwise, anyone could access your pipeline and use your secrets.

In this section, you learned how to avoid common mistakes, as well as how these errors can be mitigated. In the next section, you will learn more about following best practices.

Best practices

The importance of following best practices becomes evident when you encounter issues during a pipeline's life cycle. They will enable your team to not only maintain pipelines with ease but also swiftly identify and address any problems that may arise.

Here are some practices that will help your team find any issues easily:

- Use YAML syntax rather than the classic version. YAML can be version-controlled in repositories such as Azure Repos, GitHub, and Bitbucket, which can help you track changes in the YAML file and introduce changes gradually when combined with a branching strategy. Even better, if you want to enforce this, you can disable the creation of classic build pipelines at the organization or project settings level.

- Separate common tasks into separate YAML files that can be referenced from your main template when you have many tasks repeated throughout stages. This makes it easier to maintain or change the behavior of a pipeline in large multi-stage configurations.

- Consider creating more stages if you have long-running jobs.

- Keep your pipelines small and focused, using templates, stages, jobs, and steps.

- Build stages for large and complex applications might require more resources to compile and package faster. Microsoft-hosted agents run on general-purpose virtual machines, with only dual-core CPUs, 7 GB of RAM, and a 14 GB SSD for Linux and Windows, or three-core CPUs, 14 GB of RAM, and a 14 GB SSD for macOS. If you need to improve the performance of builds, you should consider self-hosted agents running on virtual machines with more resources.

- Consider using programming languages that support incremental compilation. Application projects that can take advantage of partial incremental builds can only be hosted in self-hosted agents, ensuring that filesystem resources are not discarded after every build.

- Use variable groups to reduce duplication when multiple pipelines use the same value. This makes your pipelines more maintainable.

- Use secret values for private values or keys. Navigate to **Library** | **secrets** | **Variable group**, and click on the lock icon to limit access to the variables, as shown in the following screenshot:

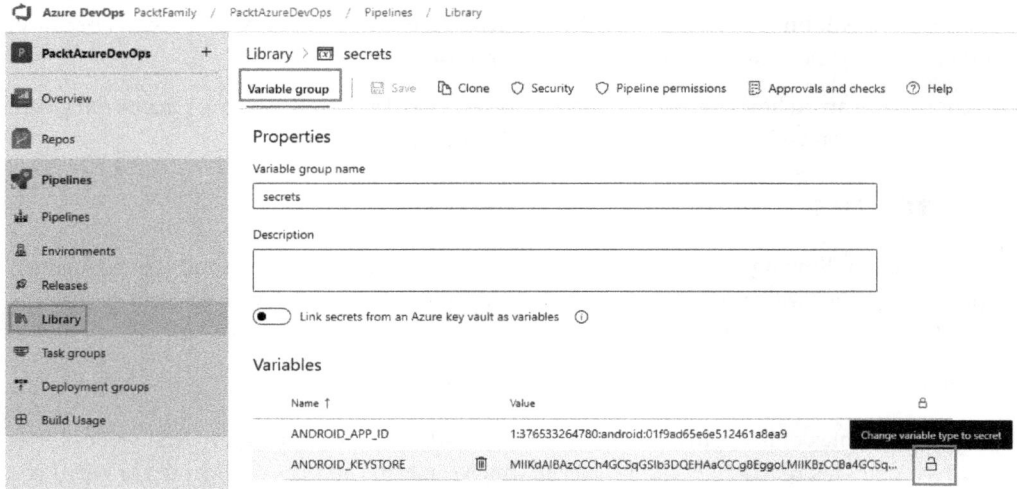

Figure 12.5 – Creating secret variables

- If you have variables from Azure **Key Vault** (**KV**), Azure's **Key Management Service** (**KMS**) will encrypt and store the secret key safely. It will make sure all variables are safe. Once this is done, no one can read your secret keys without your permission. It is recommended that you link all variables from KV to the variable group, as shown here:

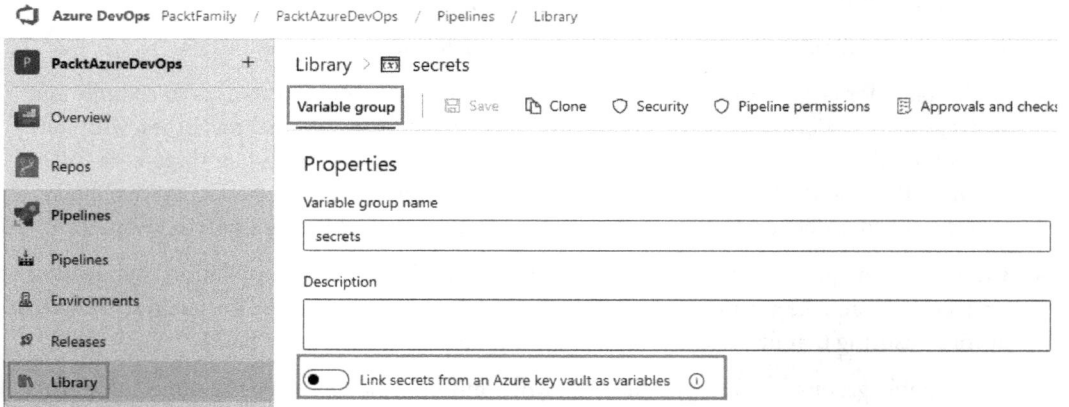

Figure 12.6 – Linking secrets from Azure KV

- Consider using the **principle of least privilege,** a security concept that dictates that a user should be granted the minimum levels of access – or permissions – needed to perform their job functions. This principle is applied to restrict users' access rights to what is only strictly required to do their work, minimizing the risk of malicious activity or accidental damage and reducing the risk of the wrong pipelines being used.

- To investigate the problem of failed pipelines, you can download the log directly from the pipelines. This can be done by following these steps:

  I.   Select the pipeline for which you need to download the log.

  II.   Click on the three dots icon and then **Download logs**:

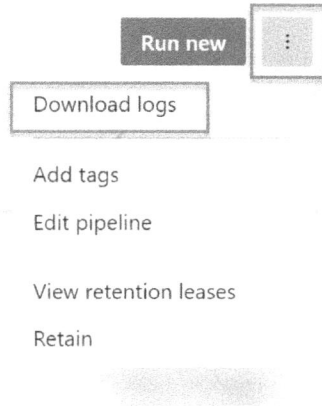

Figure 12.7 – Download logs for pipelines

- If you use `PublishBuildArtifacts@1`, you can download artifacts directly from a pipeline if you need to inspect them, by following these steps:

  I.   Select a pipeline, and click on the value in the **Related** column, **# published**:

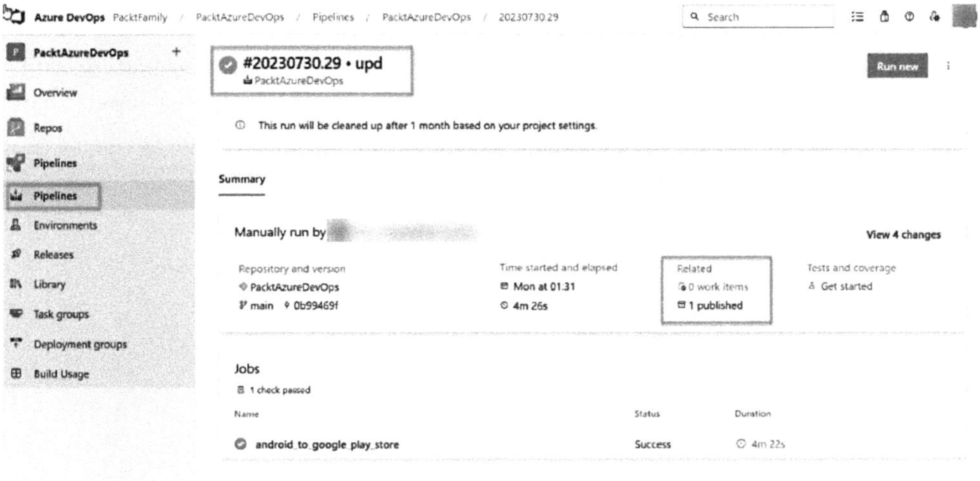

Figure 12.8 – Pipeline artifacts

II.   Click on the file to download it:

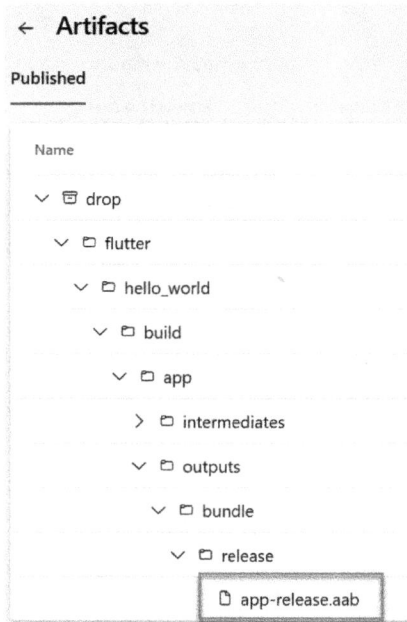

Figure 12.9 – Downloading a file from Artifacts

- For cost optimization, you can adjust the retention policy of **Artifacts** to reduce the storage size:

I.   Go to **Project Settings | Settings**.

II.   Adjust the parameters of the retention policy, depending on your business objectives. For example, if your business plan requires you to retain Azure pipeline results for only five days due to the need for cost optimization in storage, then it's essential to adhere to this timeframe. Keeping data for longer periods would result in additional storage costs:

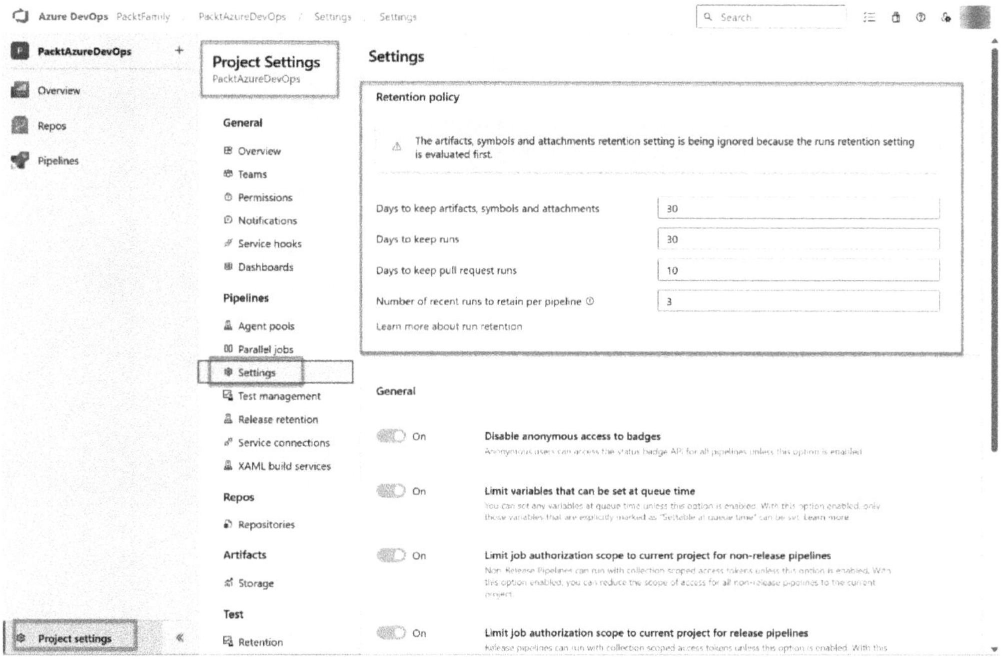

Figure 12.10 – Retention policy

- Consider validating the number of parallel jobs you have. It will help you to understand better how many processes you can run on the pipelines in parallel mode:

  I.   Navigate to **Project Settings | Parallel jobs**.

  II.   You can see an overview of all the agents you have:

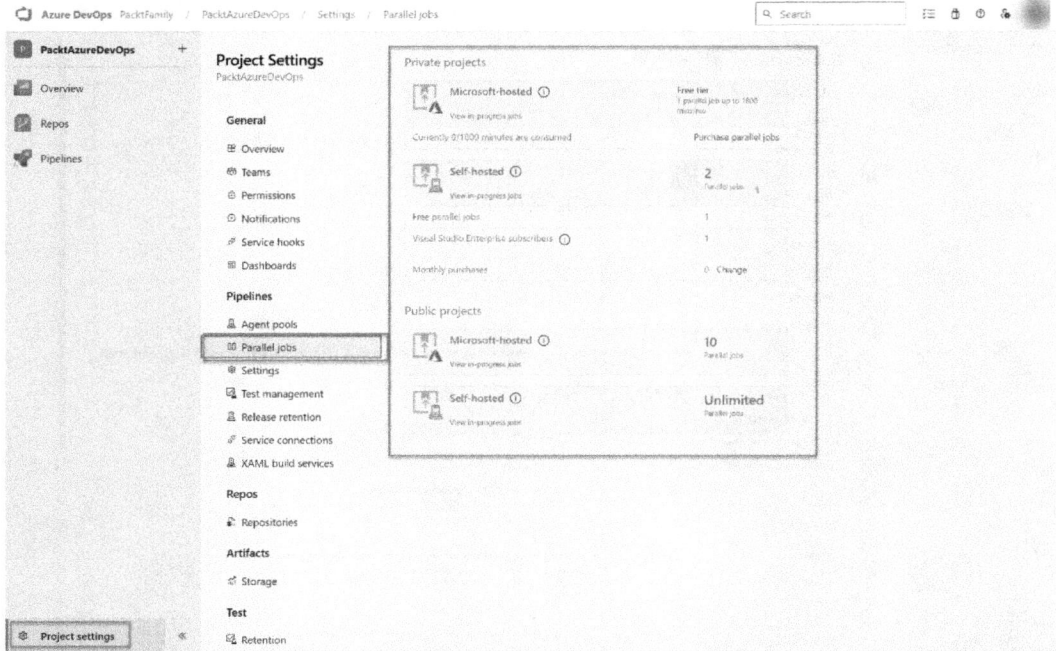

Figure 12.11 – Parallel jobs

In this section, you learned how to make a better Azure pipeline based on best practices. In the last section of this chapter, you will learn more about the future of Azure Pipelines.

Future trends

Azure DevOps is an ever-evolving product, and as such, new features and capabilities are introduced often, typically because of the Microsoft product team looking to improve the product, or because of requests from users.

Let's look at some of the most recent features related to Azure Pipelines:

- Improvements in the authoring experience for Azure Pipelines tasks to support newer versions of Node.js and ease the transition to newer versions of it, allowing you and any other contributors in the community to create your own extensions.

- GitHub Advanced Security for Azure DevOps is a new security suite of tools aimed at increasing **DevSecOps** practices. Among its features, it includes the ability to scan your code to prevent common vulnerability scenarios, using specialized tasks that detect scenarios such as cross-site scripting, SQL or XML injection attacks, and many more common issues that developers can often overlook. This is available for many programming languages and keeps evolving to offer more support in the future.

> **What is DevSecOps?**
>
> DevSecOps is a comprehensive integration of DevOps and security practices in the software development life cycle, which promotes that teams collaborate closely to include security in all aspects of the development process. This reduces and eliminates the risks associated with vulnerabilities in their code and third-party dependencies as early as possible.

- Reduce the gaps in features between YAML pipelines and classic release management deployment pipelines for CD scenarios. Several capabilities to improve the checks will be released in the coming year.

Looking into the future, there are other features that will also become important as you expand your usage of Azure Pipelines:

- In-product recommendations to highlight the best practices when configuring your pipelines around security
- Deprecation and removal of unsupported versions of Node.js for the agent software included with both the Microsoft-hosted and self-hosted versions

The Azure DevOps features roadmap is available at `https://learn.microsoft.com/en-us/azure/devops/release-notes/features-timeline`, where you can see all the product capabilities that are in progress or planned to be released by Microsoft in the near and long term. If a feature is in progress and slated to be released, you will see the approximate year and quarter when the feature will be available, as well as its availability in the cloud service or the server edition of the product. You can influence the future of the product by making suggestions through the portal, as shown in the following screenshot:

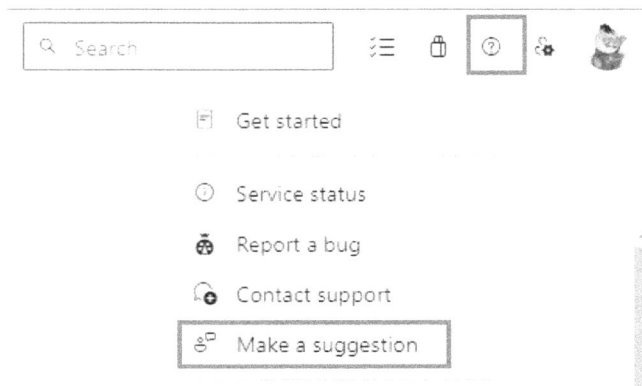

Figure 12.12 – Make a suggestion to improve Azure DevOps

Now that we've looked at some future trends of Azure Pipelines, let's wrap things up.

Summary

This chapter has been a crucial stepping stone in your journey, empowering you to use Azure Pipelines with confidence and expertise. You're now better equipped to recognize and address the common pitfalls and challenges, thereby saving valuable resources and maintaining a seamless workflow.

Following the best practices will help optimize the functionality of your pipelines and enhance collaboration within your team, ensuring the efficiency and reliability of your pipeline operations. We also offered a glimpse into the future of Azure Pipelines, which will enable you to adapt and stay ahead in this dynamic field.

Armed with the knowledge you have acquired throughout this book, you can now navigate the Azure Pipelines landscape, deploying your applications across diverse scenarios and requirements with efficiency and precision.

Good luck on your journey!

Index

‹packt›

Other Books You May Enjoy

If you enjoyed this book, you may be interested in these other books by Packt:

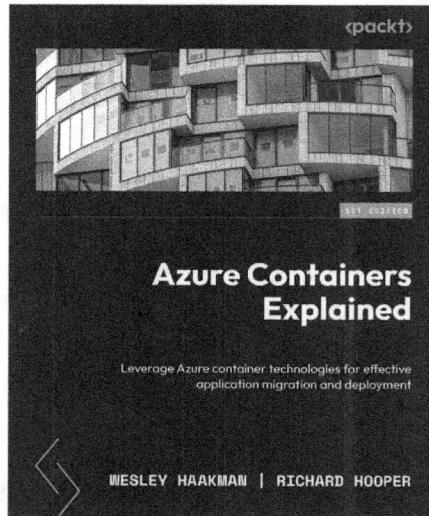

Azure Containers Explained

Wesley Haakman, Richard Hooper

ISBN: 978-1-80323-105-1

- Make the best-suited architectural choices to meet your business and application needs
- Understand the migration paths between different Azure Container services
- Deploy containerized applications on Azure to multiple technologies
- Know when to use Azure Container Apps versus Azure Kubernetes Service
- Find out how to add features to an AKS cluster
- Investigate the containers on Azure Web apps and Functions apps
- Discover ways to improve your current architecture without starting again
- Explore the financial implications of using Azure container services

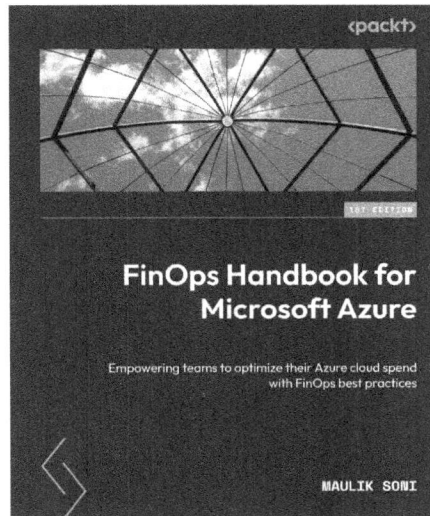

FinOps Handbook for Microsoft Azure

Maulik Soni

ISBN: 978-1-80181-016-6

- Get the grip of all the activities of FinOps phases for Microsoft Azure
- Understand architectural patterns for interruptible workload on Spot VMs
- Optimize savings with Reservations, Savings Plans, Spot VMs
- Analyze waste with customizable pre-built workbooks
- Write an effective financial business case for savings
- Apply your learning to three real-world case studies
- Forecast cloud spend, set budgets, and track accurately

Packt is searching for authors like you

If you're interested in becoming an author for Packt, please visit `authors.packtpub.com` and apply today. We have worked with thousands of developers and tech professionals, just like you, to help them share their insight with the global tech community. You can make a general application, apply for a specific hot topic that we are recruiting an author for, or submit your own idea.

Share Your Thoughts

Now you've finished *Implementing CI/CD Using Azure Pipelines*, we'd love to hear your thoughts! Scan the QR code below to go straight to the Amazon review page for this book and share your feedback or leave a review on the site that you purchased it from.

`https://packt.link/r/1804612499`

Your review is important to us and the tech community and will help us make sure we're delivering excellent quality content.

Download a free PDF copy of this book

Thanks for purchasing this book!

Do you like to read on the go but are unable to carry your print books everywhere?

Is your eBook purchase not compatible with the device of your choice?

Don't worry, now with every Packt book you get a DRM-free PDF version of that book at no cost.

Read anywhere, any place, on any device. Search, copy, and paste code from your favorite technical books directly into your application.

The perks don't stop there, you can get exclusive access to discounts, newsletters, and great free content in your inbox daily

Follow these simple steps to get the benefits:

1. Scan the QR code or visit the link below

https://packt.link/free-ebook/978-1-80461-249-1

2. Submit your proof of purchase

3. That's it! We'll send your free PDF and other benefits to your email directly

www.ingramcontent.com/pod-product-compliance
Lightning Source LLC
Chambersburg PA
CBHW081052220326
41598CB00038B/7062